FUNCTIONS OF
A COMPLEX
VARIABLE

WITH APPLICATIONS

LONGMAN MATHEMATICAL TEXTS
Edited by Alan Jeffrey and Iain Adamson

FUNCTIONS OF
A COMPLEX
VARIABLE
WITH APPLICATIONS

E. G. PHILLIPS

LONGMAN

LONGMAN GROUP LIMITED
London
Associated companies, branches and representatives
throughout the world

First published by Oliver & Boyd 1940
Eighth edition 1957
Reprinted 1961, 1963, 1965, 1968, 1972
Reprinted by Longman Group Ltd. 1975

ISBN 0 582 44286 9

Printed in Great Britain by
Whitstable Litho Ltd., Whitstable, Kent

PREFACE TO THE EIGHTH EDITION

CHANGES that have been made in recent editions include a set of Miscellaneous Examples at the end of the book and an independent proof of Liouville's theorem has been given. In this edition, the proof of the Example on page 50 has been altered.

Limitations of space made it necessary for me to confine myself to the more essential aspects of the theory and its applications, but I have aimed at including those parts of the subject which are most useful to Honours students. Many readers may desire to extend their knowledge of the subject beyond the limits of the present book. Such readers are recommended to study the standard treatises of Copson, *Functions of a Complex Variable* (Oxford, 1935), and Titchmarsh, *Theory of Functions* (Oxford, 1939). I take this opportunity of acknowledging my constant indebtedness to these works both in material and presentation.

I have presupposed a knowledge of Real Variable Theory corresponding approximately to the content of my *Course of Analysis* (Cambridge, Second Edition, 1939). References are occasionally given to this book in footnotes as P.A.

I wish to express my thanks to all those friends who have made helpful suggestions. In particular, I mention two of my colleagues, Mr A. C. Stevenson, of University College, London, who read the proofs of the first edition, and Prof. H. Davenport, F.R.S., who very kindly suggested a number of improvements for the second edition. I desire also to express my gratitude to the publishers for the careful and efficient way in which they have carried out their duties.   E. G. P.

BANGOR, *October* 1956

CONTENTS

THE CALCULUS OF RESIDUES—*continued.*

CHAPTER I

FUNCTIONS OF A COMPLEX VARIABLE

§ 1. Complex Numbers

This book is concerned essentially with the application of the methods of the differential and integral calculus to complex numbers. A number of the form $a+i\beta$, where i is $\sqrt{(-1)}$ and a and β are real numbers, is called a **complex number**; and, although complex numbers are capable of a geometrical interpretation, it is important to give a *definition* of them which depends only on real numbers. Complex numbers first became necessary in the study of algebraic equations. It is desirable to be able to say that every quadratic equation has two roots, every cubic equation has three roots, and so on. If real numbers only are considered, the equation $x^2+1 = 0$ has no roots and $x^3-1 = 0$ has only one. Every generalisation of number first presented itself as needed for some simple problem, but extensions of number are not created by the mere need of them; they are created by the definition, and our object is now to *define* complex numbers.

By choosing one of several possible lines of procedure, we *define a complex number as an ordered pair of real numbers*. Thus $(4, 3)$, $(\sqrt{2}, e)$, $(\tfrac{1}{4}, \pi)$ are complex numbers. If we write $z = (x, y)$, x is called the **real part,** and y the **imaginary part,** of the complex number z.

(i) Two complex numbers are **equal** if, and only if, their real and imaginary parts are separately equal. The equation $z = z'$ implies that both $x = x'$ and $y = y'$.

(ii) **The modulus** of z, written $|z|$, is defined to be $+\sqrt{(x^2+y^2)}$. It follows immediately from the definition that $|z| = 0$ if, and only if, $x = 0$ and $y = 0$.

(iii) *The fundamental operations.*

If $z = (x, y)$, $z' = (x', y')$ we have the following *definitions* :

(1) $z+z'$ is $(x+x', y+y')$.
(2) $-z$ is $(-x, -y)$.
(3) $z-z' = z+(-z')$ is $(x-x', y-y')$.
(4) zz' is $(xx'-yy', xy'+x'y)$.

If the fundamental operations are thus defined, we easily see that the fundamental laws of algebra are all satisfied.

(a) *The commutative and associative laws of addition hold* :

$$z_1+z_2 = z_2+z_1 ;$$
$$z_1+(z_2+z_3) = (z_1+z_2)+z_3 = z_1+z_2+z_3.$$

(b) *The same laws of multiplication hold* :

$$z_1z_2 = z_2z_1 ;$$
$$z_1(z_2z_3) = (z_1z_2)z_3 = z_1z_2z_3.$$

(c) *The distributive law holds* :

$$(z_1+z_2)z_3 = z_1z_3+z_2z_3.$$

As an example of the method, we show that the commutative law of multiplication holds. The others are proved similarly.

$$z_1z_2 = (x_1x_2-y_1y_2, x_1y_2+x_2y_1)$$
$$= (x_2x_1-y_2y_1, x_2y_1+x_1y_2) = z_2z_1.$$

We have thus seen that complex numbers, as defined above, obey the fundamental laws of the algebra of real numbers : hence their algebra will be identical in *form*, though not in *meaning*, with the algebra of real numbers.

We observe that there is no order among complex numbers. As applied to complex numbers, the phrases " greater than " or " less than " have no meaning. Inequalities can only occur in relations between the *moduli* of complex numbers.

(iv) *The definition of division.*

Consider the equation $z\zeta = z'$, where $z = (x, y)$, $\zeta = (\xi, \eta)$, $z' = (x', y')$, then we have

$$(x\xi - y\eta, \ x\eta + y\xi) = (x', y'),$$

so that $\qquad x\xi - y\eta = x', \ x\eta + y\xi = y',$

and, on solving for ξ and η,

$$\xi = \frac{yy' + xx'}{x^2 + y^2}, \ \eta = \frac{xy' - x'y}{x^2 + y^2} \ ;$$

provided that $|z| \neq 0$. Hence, if $|z| \neq 0$, there is a unique solution, and $\zeta = (\xi, \eta)$ is the quotient z'/z.

Division by a complex number whose modulus is zero is meaningless ; this conforms with the algebra of real numbers, in which division by zero is meaningless.

The abbreviated notation.

It is customary to denote a complex number whose imaginary part is zero by the real-number symbol x. If we adopt this practice, it is essential to realise that x may have two meanings (i) the real number x, and (ii) the complex number $(x, 0)$. Although in theory it is important to distinguish between (i) and (ii), in practice it is legitimate to confuse them ; and if we use the abbreviated notation, in which x stands for $(x, 0)$ and y for $(y, 0)$, then

$x + y = (x, 0) + (y, 0) = (x + y, \ 0),$
$\quad xy = (x, 0) \cdot (y, 0) = (x \cdot y - 0 \cdot 0, \ x \cdot 0 + 0 \cdot y) = (xy, 0).$

Hence, so far as sums and products are concerned, complex numbers whose imaginary parts are zero can be treated as though they were real numbers. It is customary to denote the complex number $(0, 1)$ by i. With this convention, $i^2 = (0, 1) \cdot (0, 1) = (-1, 0)$, so that i may be regarded as the square root of the real number -1. On using the abbreviated notation, it follows that

$$(x, y) = x + iy,$$

for, since $i = (0, 1)$, we have

$$x+iy = (x, 0)+(0, 1) \cdot (y, 0)$$
$$= (x, 0)+(0 \cdot y-1 \cdot 0, 0 \cdot 0+1 \cdot y)$$
$$= (x, 0)+(0, y) = (x+0, 0+y) = (x, y).$$

In virtue of this relation we see that, in any operation involving sums and products, it is allowable to treat x, y and i as though they were ordinary real numbers, with the proviso that i^2 must always be replaced by -1.

§ 2. Conjugate Complex Numbers

If $z = x+iy$, it is customary to write $x = \mathbf{R}z$, $y = \mathbf{I}z$. The number $x-iy$ is said to be **conjugate** to z and is usually denoted by $\bar{z}$. It readily follows that the numbers conjugate to z_1+z_2 and z_1z_2 are $\bar{z}_1+\bar{z}_2$ and $\bar{z}_1\bar{z}_2$ respectively.

Proofs of theorems on complex numbers are often considerably simplified by the use of conjugate complex numbers, in virtue of the relations, easily proved,

$$| z |^2 = z\bar{z}, \quad 2\mathbf{R}z = z+\bar{z}, \quad 2i\mathbf{I}z = z-\bar{z}.$$

To prove that *the modulus of the product of two complex numbers is the product of their moduli*, we proceed as follows :

$$| z_1z_2 |^2 = z_1z_2\bar{z}_1\bar{z}_2 = z_1\bar{z}_1 \cdot z_2\bar{z}_2 = | z_1 |^2 \cdot | z_2 |^2$$

and so, since the modulus of a complex number is never negative,

$$| z_1z_2 | = | z_1 | \cdot | z_2 |.$$

Theorem. The modulus of the sum of two complex numbers cannot exceed the sum of their moduli.

$$| z_1+z_2 |^2 = (z_1+z_2)(\bar{z}_1+\bar{z}_2)$$
$$= z_1\bar{z}_1+z_1\bar{z}_2+\bar{z}_1z_2+z_2\bar{z}_2$$
$$= | z_1 |^2+2\mathbf{R}(z_1\bar{z}_2)+| z_2 |^2$$
$$\leqslant | z_1 |^2+2 | z_1\bar{z}_2 | + | z_2 |^2$$
$$= (| z_1 |+| z_2 |)^2,$$

and so $| z_1+z_2 | \leqslant | z_1 |+| z_2 |$;

a result which can be readily extended by induction to any finite number of complex numbers.

In a similar way we can prove another useful result, viz.

$$|z_1-z_2| \geqslant |(|z_1|-|z_2|)|.$$

We have

$$|z_1-z_2|^2 = |z_1|^2 - 2\mathbf{R}(z_1\bar{z}_2) + |z_2|^2$$
$$\geqslant |z_1|^2 - 2|z_1\bar{z}_2| + |z_2|^2$$
$$= (|z_1|-|z_2|)^2 \,;$$

hence
$$|z_1-z_2| \geqslant |(|z_1|-|z_2|)|.$$

§ 3. Geometrical Representation of Complex Numbers

If we denote $(x^2+y^2)^{\frac{1}{2}}$ by r, and choose θ so that $r\cos\theta = x$, $r\sin\theta = y$, then r and θ are clearly the radius

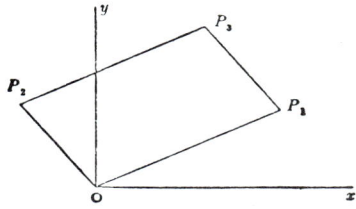

Fig. 1.

vector and vectorial angle of the point P, (x, y), referred to an origin O and rectangular axes Ox, Oy. It is clear that any complex number can be represented geometrically by the point P, whose Cartesian coordinates are (x, y) or whose polar coordinates are (r, θ), and the representation of complex numbers thus afforded is called the **Argand diagram**.

By the definition already given, it is evident that r is the modulus of $z = (x, y)$; the angle θ is called the **argument** of z, written $\theta = \arg z$. The argument is not unique, for if θ be a value of the argument, so also is

$2n\pi+\theta$, $(n = 0, \pm 1, \pm 2, ...)$. The **principal value** of $\arg z$ is that which satisfies the inequalities $-\pi < \arg z \leqslant \pi$.

Let P_1 and P_2 (in fig. 1) be the points z_1 and z_2, then we can represent *addition* in the following way. Through P_1, draw P_1P_3 equal to, and parallel to OP_2. Then P_3 has coordinates (x_1+x_2, y_1+y_2) and so P_3 represents the point z_1+z_2.

In vectorial notation,

$$\overline{OP_3} = \overline{OP_1}+\overline{P_1P_3} = \overline{OP_1}+\overline{OP_2} = \overline{OP_2}+\overline{P_2P_3}.$$

Similarly, we have, if P_3 is the point z_3,

$$z_3-z_2 = \overline{OP_3}-\overline{OP_2} = \overline{OP_3}+\overline{P_3P_1} = \overline{OP_1} = z_1.$$

It is convenient to write $\operatorname{cis}\theta$ for $\cos\theta+i\sin\theta$. If $z_1 = r_1\operatorname{cis}\theta_1$, $z_2 = r_2\operatorname{cis}\theta_2$, ..., $z_n = r_n\operatorname{cis}\theta_n$, then, by de Moivre's theorem,

$$z_1z_2...z_n = r_1r_2...r_n\operatorname{cis}(\theta_1+\theta_2+...+\theta_n),$$

which readily exhibits the fact that the modulus and argument of a product are equal respectively to the product of the moduli and the sum of the arguments of the factors. In particular, if n be a positive integer and $z = r\operatorname{cis}\theta$, $z^n = r^n\operatorname{cis} n\theta$.

Similarly,

$$\frac{z_1}{z_2} = \frac{r_1}{r_2}\operatorname{cis}(\theta_1-\theta_2).$$

If n is a positive integer, there are n distinct values of $z^{1/n}$. If m is any integer, since

$$\left(\operatorname{cis}\frac{\theta+2m\pi}{n}\right)^n = \operatorname{cis}\theta,$$

it follows that $r^{1/n}\operatorname{cis}\{(\theta+2m\pi)/n\}$ is an nth root of $z=r\operatorname{cis}\theta$. If we substitute the numbers $0, 1, 2, ... n-1$ in succession for m, we obtain n distinct values of $z^{1/n}$; and the substitution of other integers for m merely gives rise to repetitions of these values. Also, there can be no other values, since $z^{1/n}$ is a root of the equation $u^n = z$ which cannot have more than n roots.

Similarly, if p and q are integers prime to each other and q is positive,

$$z^{p/q} = r^{p/q} \operatorname{cis}\{(p\theta + 2m\pi)/q\},$$

where $m = 0, 1, 2, ..., q-1$.

By considering the modulus and argument of a complex number, the operation of multiplying any complex number $x+iy$ by i is easily seen to be equivalent to turning the line OP through a right-angle in the positive (counter-clockwise) sense. We have just seen that

$$\arg(z_1 z_2) = \arg z_1 + \arg z_2, \quad \arg\left(\frac{z_1}{z_2}\right) = \arg z_1 - \arg z_2,$$

so that the formal process of " taking arguments " is similar to that of " taking logarithms." Hence, if $\arg(x+iy) = a$,

$$\arg i(x+iy) = \arg i + \arg(x+iy) = \tfrac{1}{2}\pi + a.$$

Since $|i| = 1$, multiplying by i leaves $|x+iy|$ unaltered.

§ 4. Sets of Points in the Argand Diagram

We now explain some of the terminology necessary for dealing with sets of complex numbers in the Argand diagram. We shall use such terms as *domain, contour, inside and outside of a closed contour*, without more precise definition than geometrical intuition requires. The general study of such questions as the precise determination of the *inside* and *outside* of a closed contour is not so easy as our intuitions might lead us to expect.* For our present purpose, however, we shall find that no difficulties arise from our relying upon geometrical intuition.

By a **neighbourhood** of a point z_0 in the Argand diagram, we mean the set of all points z such that $|z-z_0| < \epsilon$, where ϵ is a given positive number. A point z_0 is said

* For further information, see *e.g.* Dienes, *The Taylor Series* (Oxford, 1931), Ch. VI.

to be a **limit point** of a set of points S, if every neighbour-hood of z_0 contains a point of S other than z_0. The definition implies that every neighbourhood of a limit point z_0 contains an infinite number of points of S. For, the neighbourhood $|z-z_0|<\epsilon$ contains a point z_1 of S distinct from z_0, the neighbourhood $|z-z_0|<|z_1-z_0|$ contains a point z_2 of S distinct from z_0 and so on indefinitely.

The limit points of a set are not necessarily points of the set. If, however, every limit point of the set belongs to the set, we say that the set is **closed**. There are two types of limit points, *interior points* and *boundary points*. A limit point z_0 of S is an **interior point** if there exists a neighbourhood of z_0 which consists entirely of points of S. A limit point which is not an interior point is a **boundary point**.

A set which consists entirely of interior points is said to be an **open set**.

It should be observed that a set need not be either open or closed. An example of such a set is that consisting of the point $z = 1$ and all the points for which $|z|<1$.

We now define a *Jordan curve*.

The equation $z = x(t)+iy(t)$, where $x(t)$ and $y(t)$ are real continuous functions of the real variable t, defined in the range $\alpha \leqslant t \leqslant \beta$, determines a set of points in the Argand diagram which is called a **continuous arc**. A point z_1 is a **multiple point** of the arc, if the equation $z_1 = x(t)+iy(t)$ is satisfied by more than one value of t in the given range.

A continuous arc without multiple points is called a **Jordan arc**. If the points corresponding to the values α and β coincide, the arc, which has only one multiple point, a double point corresponding to the terminal values α and β of t, is called a **simple closed Jordan curve**.

A set of points is said to be **bounded** if there exists a constant K such that $|z|\leqslant K$ is satisfied for all points

z of the set.   If no such number K exists the set is **unbounded.**

A **domain** is defined as follows:—

A set of points in the Argand diagram is said to be **connex** if every pair of its points can be joined by a polygonal arc which consists only of points of the set. An **open domain** is an open connex set of points.   The set, obtained by adding to an open domain its boundary points, is called a *closed domain.*

The *Jordan curve theorem* states that *a simple closed Jordan curve divides the plane into two open domains which have the curve as common boundary.*   Of these domains one is bounded and it is called the **interior,** the other, which is unbounded, is called the **exterior.**   Although the result stated seems quite obvious, the proof is very complicated and difficult.   When using simple closed Jordan curves consisting of a few straight lines and circular arcs, geometrical intuition makes it obvious which is the interior and which is the exterior domain.

For example, the circle $|z| = R$ divides the Argand diagram into two separated open domains $|z| < R$ and $|z| > R$.   Of these the former is a bounded domain and is the interior of the circle $|z| = R$;   the latter, which is unbounded, is the exterior of the circle $|z| = R$.

In complex variable theory we complete the complex plane by adding a *single* **point at infinity.**   This point is defined to be the point corresponding to the origin by the transformation $z' = 1/z$.

§ 5. Functions of a Complex Variable.   Continuity

If $w(= u + iv)$ and $z(= x + iy)$ are any two complex numbers, we might say that w is a function of z, $w = f(z)$, if, to every value of z in a certain domain D, there correspond one or more values of w.   This definition, similar to that given for real variables, is quite legitimate, but it is futile because it is too wide.   On this definition, a function of the complex variable z is exactly the same thing as a complex function

$$u(x, y) + iv(x, y)$$

of two real variables x and y.

For functions defined in this way, the definition of continuity is exactly the same as that for functions of a real variable. The function $f(z)$ is **continuous** at the point z_0 if, given any $\epsilon, > 0$, we can find a number δ such that

$$|f(z) - f(z_0)| < \epsilon$$

for all points z of D satisfying $|z - z_0| < \delta$. The number δ depends on ϵ and also, in general, upon z_0. If it is possible to find a number $h(\epsilon)$ *independent of z_0*, such that $|f(z) - f(z_0)| < \epsilon$ holds for every pair of points z, z_0 of the domain D for which $|z - z_0| < h$, then $f(z)$ is said to be **uniformly** continuous in D. It can be proved that a function which is continuous in a bounded closed domain is uniformly continuous there.*

It is easy to show that this definition of continuity is equivalent to the statement that a continuous function of z is merely a continuous complex function of the two variables x and y, for, if

$$f(z) = u(x, y) + iv(x, y),$$

when $f(z)$ is continuous on the above definition, so are $u(x, y)$ and $v(x, y)$; and conversely, if u and v are continuous functions of x and y, $f(z)$ is a continuous function of z.

The only class of functions of z which is of any practical utility is the class of functions to which the process of differentiation can be applied.

§ 6. Differentiability

We next consider whether the definition of the derivative of a function of a single real variable is applicable

---

* For a proof of this theorem for a closed interval, see Phillips, *A Course of Analysis* (Cambridge, 1939), p. 73. This will be referred to subsequently as P.A.

to functions of a complex variable. The natural definition is as follows: *Let $f(z)$ be a one-valued function, defined in a domain D of the Argand diagram, then $f(z)$ is differentiable at a point z_0 of D if*

$$\frac{f(z) - f(z_0)}{z - z_0}$$

tends to a unique limit as $z \to z_0$, provided that z is also a point of D.

If the above limit exists it is called the **derivative** of $f(z)$ at $z = z_0$ and is denoted by $f'(z_0)$. Restating the definition in a more elementary form, it asserts that, given $\epsilon > 0$, we can find a number δ such that

$$\left| \frac{f(z) - f(z_0)}{z - z_0} - f'(z_0) \right| < \epsilon$$

for all z, z_0 in D satisfying $0 < |z - z_0| < \delta$. That continuity does not imply differentiability is seen from the following simple example :—

Let $f(z) = |z|^2$. This continuous function is differentiable at the origin, but nowhere else. For if $z_0 \neq 0$ we have

$$\frac{|z|^2 - |z_0|^2}{z - z_0} = \frac{z\bar{z} - z_0\bar{z}_0}{z - z_0} = \bar{z} + z_0 \frac{\bar{z} - \bar{z}_0}{z - z_0}$$
$$= \bar{z} + z_0(\cos 2\phi - i \sin 2\phi)$$

where $\phi = \arg(z - z_0)$. It is clear that this expression does not tend to a unique limit as $z \to z_0$.

If $z_0 = 0$ the incrementary ratio is $\bar{z}$, which tends to zero as $z \to 0$.

§ 7. Regular Functions

A function of z which is one-valued and differentiable at every point of a domain D is said to be **regular** * in the domain D. A function may be differentiable in a domain save possibly for a finite number of points. These points are called **singularities** of $f(z)$. We next discuss the necessary and sufficient conditions for a function to be regular.

* The terms *analytic* and *holomorphic* are sometimes used as synonymous with the term *regular* as defined above.

(1) *The necessary conditions for f(z) to be regular.*

If $f(z) = u(x, y) + iv(x, y)$ is differentiable at a given point z, the ratio $\{f(z+\Delta z)-f(z)\}/\Delta z$ must tend to a definite limit as $\Delta z \to 0$ *in any manner.* Now $\Delta z = \Delta x + i\Delta y$. Take Δz to be wholly real, so that $\Delta y = 0$, then

$$\frac{u(x+\Delta x, y)-u(x, y)}{\Delta x} + i\, \frac{v(x+\Delta x, y)-v(x, y)}{\Delta x}$$

must tend to a definite limit as $\Delta x \to 0$. It follows that the partial derivatives u_x, v_x must exist at the point (x, y) and the limit is $u_x + iv_x$. Similarly, if we take Δz to be wholly imaginary, so that $\Delta x = 0$, we find that u_y, v_y must exist at (x, y) and the limit in this case is $v_y - iu_y$. Since the two limits obtained must be identical, on equating real and imaginary parts, we get

$$u_x = v_y \,,\; u_y = -v_x. \qquad . \qquad . \qquad (1)$$

These two relations are called the **Cauchy–Riemann differential equations.**

We have thus proved that *for the function f(z) to be differentiable at the point z it is necessary that the four partial derivatives u_x, v_x, u_y, v_y should exist and satisfy the Cauchy-Riemann differential equations.*

We thus see that the results of assuming differentiability are more far-reaching than those of assuming continuity. Not only must the functions u and v possess partial derivatives of the first order, but these must be connected by the differential equations (1).

That the above conditions are necessary, but not sufficient, may be seen by considering Examples 6 and 7 at the end of this chapter.

(2) *Sufficient conditions for f(z) to be regular.*

Theorem. The continuous one-valued function f(z) is regular in a domain D if the four partial derivatives u_x, v_x, u_y, v_y exist, are continuous *and satisfy the Cauchy-Riemann equations at each point of D.*

Now

$$\Delta u = u(x+\Delta x, y+\Delta y) - u(x, y),$$
$$= u(x+\Delta x, y+\Delta y) - u(x+\Delta x, y) + u(x+\Delta x, y) - u(x, y),$$
$$= \Delta y \,.\, u_y(x+\Delta x, y+\theta\Delta y) + \Delta x \,.\, u_x(x+\theta'\Delta x, y) \,;$$

where $0<\theta<1$, $0<\theta'<1$, by the mean-value theorem.*
Since u_x, u_y are both continuous, we may write

$$\Delta u = \Delta x\{u_x(x, y)+\epsilon\} + \Delta y\{u_y(x, y)+\epsilon'\},$$

where ϵ and ϵ' both tend to zero as $|\Delta z| \to 0$.
   Similarly,

$$\Delta v = \Delta x\{v_x(x, y)+\eta\} + \Delta y\{v_y(x, y)+\eta'\},$$

where η and η' both tend to zero as $|\Delta z| \to 0$.

Hence $\Delta w = \Delta u + i\Delta v$
$$= \Delta x(u_x + iv_x) + \Delta y(u_y + iv_y) + \omega\Delta x + \omega'\Delta y,$$

where ω and ω' tend to zero as $|\Delta z| \to 0$.
   On using the Cauchy-Riemann equations we get

$$\Delta w = (\Delta x + i\Delta y)(u_x + iv_x) + \omega\Delta x + \omega'\Delta y$$

and, on dividing by Δz and taking the limit as $|\Delta z| \to 0$,

$$\frac{dw}{dz} = u_x + iv_x,$$

since

$$\left| \frac{\omega\Delta x + \omega'\Delta y}{\Delta z} \right| \leqslant |\omega| + |\omega'|\,.$$

We notice that the above sufficient conditions for the regularity of $f(z)$ require the *continuity* of the four first partial derivatives of u and v.

   If $w = u + iv$, where u and v are functions of x and y, since

$$x = \frac{1}{2}\,(z+\bar{z})\,,\; y = \frac{1}{2i}\,(z-\bar{z}),$$

* See P.A., p. 101.

u and v may be regarded formally as functions of two independent variables z and $\bar{z}$. If u and v have continuous first-order partial derivatives with respect to x and y, the condition that w shall be independent of $\bar{z}$ is that $\partial w/\partial \bar{z} = 0$. This leads to the result

$$\frac{\partial u}{\partial x}\frac{\partial x}{\partial \bar{z}} + \frac{\partial u}{\partial y}\frac{\partial y}{\partial \bar{z}} + i\left(\frac{\partial v}{\partial x}\frac{\partial x}{\partial \bar{z}} + \frac{\partial v}{\partial y}\frac{\partial y}{\partial \bar{z}}\right) = 0,$$

that is

$$\frac{1}{2}\frac{\partial u}{\partial x} - \frac{1}{2i}\frac{\partial u}{\partial y} + \frac{i}{2}\frac{\partial v}{\partial x} - \frac{1}{2}\frac{\partial v}{\partial y} = 0 ;$$

and, on writing i for $-1/i$ and equating real and imaginary parts, we get

$$\frac{\partial u}{\partial x} = \frac{\partial v}{\partial y}, \; \frac{\partial v}{\partial x} = -\frac{\partial u}{\partial y} ; \quad . \qquad . \qquad . \quad (1)$$

which are the Cauchy-Riemann equations.

Hence, in any analytical formula which represents a regular function of z, x and y can occur only in the combination $x+iy$. For example, it is clear at a glance that

$$\sin (x+3iy) = \sin (2z-\bar{z})$$

cannot be a regular function.

If $u+iv = f(x+iy)$ where $f(z)$ is a regular function, then the real functions u and v of the two real variables x and y are called **conjugate functions**.

Since the partial derivatives of u and v are connected by the relations

$$\frac{\partial u}{\partial x} = \frac{\partial v}{\partial y}, \; \frac{\partial v}{\partial x} = -\frac{\partial u}{\partial y} ; \qquad . \qquad . \quad (1)$$

if the derivatives concerned are assumed to exist and satisfy the relation $\phi_{xy} = \phi_{yx}$, it follows by partial differentiation that

$$\frac{\partial^2 v}{\partial x\partial y} = \frac{\partial^2 u}{\partial x^2} = -\frac{\partial^2 u}{\partial y^2} \text{ and } \frac{\partial^2 u}{\partial x\partial y} = -\frac{\partial^2 v}{\partial x^2} = \frac{\partial^2 v}{\partial y^2}.$$

Hence both u and v satisfy Laplace's equation in **two** dimensions

$$\nabla^2\phi \equiv \frac{\partial^2\phi}{\partial x^2} + \frac{\partial^2\phi}{\partial y^2} = 0.$$

This equation occurs frequently in mathematical physics. It is satisfied by the potential at a point not occupied by matter in a two-dimensional gravitational field. It is also satisfied by the velocity potential and stream function of two-dimensional irrotational flow of an incompressible non-viscous fluid.

By separating any regular function of z into its real and imaginary parts, we obtain immediately two solutions of Laplace's equation. It follows that the theory of functions of a complex variable has important applications to the solution of two-dimensional problems in mathematical physics. It also follows from equations (1) that

$$\frac{\partial u}{\partial x}\frac{\partial v}{\partial x} + \frac{\partial u}{\partial y}\frac{\partial v}{\partial y} = 0. \quad . \qquad . \qquad . \quad (2)$$

The geometrical interpretation of (2) is that the families of curves in the (x, y)-plane, corresponding to constant values of u and v, intersect at right angles at all their points of intersection. For if $u(x, y) = c_1$, then $du = 0$, and so

$$\frac{\partial u}{\partial x}\, dx + \frac{\partial u}{\partial y}\, dy = 0. \quad . \qquad . \qquad . \quad (3)$$

Similarly, if $v(x, y) = c_2$, we have

$$\frac{\partial v}{\partial x}\, dx + \frac{\partial v}{\partial y}\, dy = 0. \quad . \qquad . \qquad . \quad (4)$$

The condition that these families of curves intersect at right angles is

$$\left(\frac{dy}{dx}\right)_1 \left(\frac{dy}{dx}\right)_2 + 1 = 0 \quad . \qquad . \qquad . \quad (5)$$

where the suffixes 1 and 2 refer to the u and v families respectively. On using (3) and (4), it is easy to see that (5) reduces to (2).

It is possible to construct a function $f(z)$ which has a given real function of x and y for its real or imaginary part, if either of the given functions $u(x, y)$ or $v(x, y)$ is a simple combination of elementary functions satisfying Laplace's equation. A very elegant method of doing this is due to Milne-Thomson.*

Since $$x = \frac{1}{2}(z+\bar{z}), \; y = \frac{1}{2i}(z-\bar{z}),$$

$$f(z) = u\left\{\frac{z+\bar{z}}{2}, \frac{z-\bar{z}}{2i}\right\} + iv\left\{\frac{z+\bar{z}}{2}, \frac{z-\bar{z}}{2i}\right\}.$$

We can look upon this as a formal identity in two independent variables $z, \bar{z}$. On putting $\bar{z} = z$ we get

$$f(z) = u(z, 0) + iv(z, 0).$$

Now $f'(z) = u_x + iv_x = u_x - iu_y$ by the Cauchy-Riemann equations. Hence, if we write $\phi_1(x, y)$ and $\phi_2(x, y)$ for u_x and u_y respectively, we have

$$f'(z) = \phi_1(x, y) - i\phi_2(x, y) = \phi_1(z, 0) - i\phi_2(z, 0).$$

On integrating, we have

$$f(z) = \int\{\phi_1(z, 0) - i\phi_2(z, 0)\}dz + C,$$

where C is an arbitrary constant.

Similarly, if $v(x, y)$ is given, we can prove that

$$f(z) = \int\{\psi_1(z, 0) + i\psi_2(z, 0)\}dz + C,$$

where $\psi_1(x, y) = v_y$ and $\psi_2(x, y) = v_x$.

* *Math. Gazette*, **xxi.** (1937), p. 228. See also *Misc. Ex.* 1, p. 138.

As an example, suppose that $u(x,y) = e^x(x \cos y - y \sin y)$.

Here $\phi_1 = \dfrac{\partial u}{\partial x} = e^x(x \cos y - y \sin y + \cos y)$,

$\phi_2 = \dfrac{\partial u}{\partial y} = e^x(-x \sin y - \sin y - y \cos y)$.

Hence $f'(z) = \phi_1(z, 0) - i\phi_2(z, 0) = e^z (z+1)$,

and so $f(z) = \displaystyle\int e^z (z+1)dz + C = ze^z + C$.

§ 8. Power Series.   The Elementary Functions

Consider the series $\displaystyle\sum_{n=0}^{\infty} a_n z^n$ or $\displaystyle\sum_{n=0}^{\infty} a_n(z-z_0)^n$, where the coefficients a_n and z, z_0 may be complex. Since the latter series may be obtained from the former by a simple change of origin, the former may be regarded as a typical power series. It is assumed that the reader is already familiar with the theory of real power series.[*]

So far as absolute convergence is concerned, everything that has been proved for *absolutely* convergent series of *real* terms extends at once to complex series, for the series of moduli

$$|a_0| + |a_1||z| + |a_2||z|^2 + \dots$$

is a series of positive terms. The most useful convergence test for power series is Cauchy's root test, which states that a series of positive terms Σu_n is convergent or divergent according as $\overline{\lim} (u_n)^{1/n}$ is less than or greater than unity.[†] If we write $\overline{\lim} |a_n|^{1/n} = 1/R$, then we easily see that the power series $\Sigma a_n z^n$ is absolutely convergent if $|z| < R$, divergent if $|z| > R$, and if $|z| = R$ we can give no general verdict and the behaviour of the series may be of the most diverse nature. The number R is called the **radius of convergence**, and the circle, centre the origin, and radius R, is called the **circle of**

* See P.A., Ch. XIII.
† See P.A., p. 124.

convergence of the power series. Clearly there are three cases to consider (i) $R = 0$, (ii) R finite, (iii) R infinite. The first case is trivial, since the series is then convergent only when $z = 0$. In the third case the series converges for all values of z. In the second case the radius of the circle of convergence is finite and the power series is absolutely convergent at all points within this circle, and divergent at all points outside it.

We now prove an important theorem.

If $f(z) = \sum\limits_{0}^{\infty} a_n z^n$, *then the sum-function* $f(z)$ *is a regular function at every point within the circle of convergence of the power series.*

Suppose that $\Sigma a_n z^n$ is convergent for $|z| < R$. Then, if $0 < \rho < R$, $a_n \rho^n$ is bounded, say $|a_n \rho^n| < K$. Let

$$\phi(z) = \sum\limits_{n=1}^{\infty} n a_n z^{n-1}.$$

We write, for convenience, $|z| = r$, $|h| = \eta$: then, if $r < \rho$ and $r + \eta < \rho$,

$$\frac{f(z+h) - f(z)}{h} - \phi(z) = \sum\limits_{n=0}^{\infty} \left[a_n \left\{ \frac{(z+h)^n - z^n}{h} - n z^{n-1} \right\} \right].$$

Now

$$\left| \frac{(z+h)^n - z^n}{h} - n z^{n-1} \right| = \left| \frac{n(n-1)}{1 \cdot 2} z^{n-2} h + \ldots + h^{n-1} \right|$$

$$\leqslant \frac{n(n-1)}{1 \cdot 2} r^{n-2} \eta + \ldots + \eta^{n-1} = \frac{(r+\eta)^n - r^n}{\eta} - n r^{n-1}.$$

Hence

$$\left| \frac{f(z+h) - f(z)}{h} - \phi(z) \right| \leqslant K \sum\limits_{n=0}^{\infty} \frac{1}{\rho^n} \left\{ \frac{(r+\eta)^n - r^n}{\eta} - n r^{n-1} \right\}$$

$$= K \left\{ \frac{1}{\eta} \left(\frac{\rho}{\rho - r - \eta} - \frac{\rho}{\rho - r} \right) - \frac{\rho}{(\rho - r)^2} \right\}$$

$$= \frac{K \rho \eta}{(\rho - r - \eta)(\rho - r)^2},$$

which tends to zero as $\eta \to 0$.   Hence $f(z)$ has the derivative $\phi(z)$.   This proves that $f(z)$, which is plainly one-valued, is also differentiable : hence $f(z)$ is *regular* within $|z| = R$.

Since $n^{1/n} \to 1$ as $n \to \infty$, $\overline{\lim}|na_n|^{1/n} = \overline{\lim}|a_n|^{1/n} = 1/R$, and so the series $\phi(z) = \sum_{n=1}^{\infty} na_n z^{n-1}$ has the same radius of convergence as the original series.   Thus, if $\phi(z) = f'(z)$ is regular in $|z| < R$, we can show similarly that its derivative is $\sum n(n-1)a_n z^{n-2}$, and so on.   In other words, we thus prove that *a power series can be differentiated term by term as often as we please at any point within its circle of convergence.*

The above theorem, which is the analogue of a well-known theorem in real variable theory, can be superseded by a more general theorem which is one of the characteristic achievements of complex variable theory.   This theorem is as follows :

Let $f(z) = u_1(z) + u_2(z) + \ldots + u_n(z) + \ldots ;$

if each term $u_n(z)$ is regular within a region D and if the series is uniformly convergent throughout every region D' interior to D, then $f(z)$ is regular within D and all its derivatives may be calculated by term-by-term differentiation.

For a proof of this theorem, the reader is referred to larger treatises on complex variable theory.   The simpler theorem proved above will suffice for the purposes of this book.

In a later chapter (§ **34**) we prove Taylor's theorem that a function $f(z)$ can be expanded in a power series $\sum_{n=0}^{\infty} a_n(z-a)^n$ about any point a, provided that $f(z)$ is regular in $|z-a| \leqslant \rho$.   By combining Taylor's theorem with the theorem proved above, we see that the necessary and sufficient condition that a function $f(z)$ may be expanded in a power series is that it should be regular in a region. The Weierstrassian development of complex variable theory begins by *defining* an " analytic function " of z as a function expansible in a power series.   (See § **39**.)

We now consider briefly the definitions of the so-called **elementary functions** of a complex variable.

I. *Rational functions.*

A polynomial in z, $a_0 + a_1 z + \ldots + a_m z^m$, may be regarded as a power series which converges for all values of z. Since such functions are regular in the whole plane, rational functions of the type

$$f(z) = \frac{a_0 + a_1 z + \ldots + a_m z^m}{b_0 + b_1 z + \ldots + b_k z^k},$$

are regular at all points of the plane at which the denominator does not vanish. If we choose a point z_0, at which the denominator does not vanish, and replace z by $z_0 + (z - z_0)$, the function $f(z)$ becomes

$$\frac{A_0 + A_1(z - z_0) + \ldots + A_m(z - z_0)^m}{B_0 + B_1(z - z_0) + \ldots + B_k(z - z_0)^k},$$

in which $B_0 \neq 0$. It readily follows that $f(z)$ may be expanded in a power series of the form $\sum\limits_0^\infty c_n (z - z_0)^n$.

II. *The exponential function.*

For the exponential function of a *real* variable, one method of development is to define $\exp x$ as the sum-function of the power series

$$1 + x + \frac{x^2}{2!} + \frac{x^3}{3!} + \ldots . \quad . \quad . \quad (1)$$

and, on using the multiplication theorem for absolutely convergent series, we prove that *

$$\exp x \cdot \exp x' = \exp (x + x').$$

* See P.A., p. 246.

In the same way we can define $\exp z$ as the sum-function of the series of complex terms

$$1+z+ \frac{z^2}{2!} + \frac{z^3}{3!} +\cdots.$$

Since the series converges for all values of z, it defines a function regular in the whole z-plane. Such functions are called **integral functions**.

When x is rational, $\exp x$ is identical with the function e^x of elementary algebra, and when x is irrational we *define* e^x to be identical with the function $\exp x$, the sum-function of the power series (1) above. In the same way, when z is complex, we find it convenient conventionally to write e^z for $\exp z$. Since the formula $\exp z . \exp \zeta = \exp(z+\zeta)$ can be proved by multiplication of series, whether z be real or complex, the real number e with a *complex* exponent obeys the formal law of indices of elementary algebra

$$e^z. e^\zeta = e^{z+\zeta}.$$

Thus we may define the power e^z, without ambiguity, by the equation

$$e^z = 1+z+ \frac{z^2}{2!} + \frac{z^3}{3!} +\cdots;$$

and, if a is any positive number, a^z denotes the value unambiguously determined by the formula

$$a^z = e^{z \log a},$$

where $\log a$ is the real natural logarithm of a.

The reader should notice how far this definition is removed from the elementary definition " x^k is the product of k factors equal to x." At first sight there is no knowing what value belongs to a number of the form 2^i, but its value is *uniquely* determined by our definition.

For real values of y we have

$$e^{iy} = \sum_{n=0}^{\infty} \frac{(iy)^n}{n!} = \sum_{k=0}^{\infty} (-1)^k \frac{y^{2k}}{(2k)!} + i\sum_{k=0}^{\infty} (-1)^k \frac{y^{2k+1}}{(2k+1)!} \quad (2)$$
$$= \cos y + i \sin y \,;$$

since the cosine and sine of the real variable y are *defined* by the two power series on the right of (2).

Hence $e^z = e^{x+iy} = e^x e^{iy} = e^x \operatorname{cis} y.$

We also see that, since

$$|e^{iy}| = |\operatorname{cis} y| = 1 \;,\; |e^z| = |e^x||e^{iy}| = e^x,$$

since $e^x > 0$. Similarly, $\arg e^z = \mathrm{I}z = y$. *The function e^z has the period $2\pi i$*; in other words, if k is any positive or negative integer, or zero,

$$e^z = e^{z+2\pi i} = e^{z+2k\pi i},$$

for, when we increase z by $2\pi i$, y increases by 2π and this leaves the values of $\sin y$ and $\cos y$ unchanged. Every value which e^z is able to assume is therefore taken in the infinite strip $-\pi < y \leqslant \pi$, or in any strip obtainable from this by a parallel translation.

It is easy to show that e^z has no other period. If $e^z = e^\zeta$, this necessarily implies that $z = \zeta + 2k\pi i$. This follows at once, because $e^{z-\zeta} = 1$ and so

$$e^{x-\xi} \operatorname{cis}(y-\eta) = 1.$$

Hence $x-\xi = 0$, $\cos(y-\eta) = 1$, $\sin(y-\eta) = 0$; and this leads to $y-\eta = 2k\pi$, so that $z-\zeta = 2k\pi i$.

Finally, e^z never vanishes, for $e^{z_1} e^{-z_1} = 1$, and, if $e^{z_1} = 0$, this equation would give an infinite value for e^{-z_1}, which is impossible.

Since $z = x+iy = r\cos\theta + ir\sin\theta$, any complex number may be written in the form $z = re^{i\theta}$, where $|z| = r$, $\arg z = \theta$, since we have now assigned a meaning to $e^{i\theta}$.

By term-by-term differentiation of the power series defining e^z, we readily see that

$$\frac{d}{dz} e^z = e^z.$$

III. *The trigonometrical and hyperbolic functions.*

We define $\sin z$ and $\cos z$, when z is complex, as the sum-functions of power series, just as we do for $\sin x$ and $\cos x$ when x is real. Thus

$$\sin z = \sum_{n=0}^{\infty}(-1)^n \frac{z^{2n+1}}{(2n+1)!}, \ \cos z = \sum_{n=0}^{\infty}(-1)^n \frac{z^{2n}}{(2n)!};$$

and, since each of these power series has an infinite radius of convergence, $\sin z$ and $\cos z$ are integral functions.

By term-by-term differentiation of these power series, we deduce at once that the derivatives of $\sin z$ and $\cos z$ are $\cos z$ and $-\sin z$ respectively.

The other trigonometrical functions are then defined by

$$\tan z = \frac{\sin z}{\cos z}, \ \cot z = \frac{1}{\tan z}, \ \sec z = \frac{1}{\cos z}, \ \text{cosec } z = \frac{1}{\sin z}.$$

If we denote $\exp iz$ by e^{iz}, according to our agreed convention, we readily obtain the results

$$\cos z + i \sin z = e^{iz}, \ \cos z - i \sin z = e^{-iz};$$

leading to Euler's formulae

$$\cos z = \frac{1}{2}(e^{iz} + e^{-iz}), \ \sin z = \frac{1}{2i}(e^{iz} - e^{-iz}).$$

From these formulae, and the addition formula for e^z, we find that
$$\sin^2 z + \cos^2 z = 1;$$

and the addition theorems

$$\sin(z \pm \zeta) = \sin z \cos \zeta \pm \cos z \sin \zeta,$$
$$\cos(z \pm \zeta) = \cos z \cos \zeta \mp \sin z \sin \zeta,$$

also hold for complex variables. As all the elementary identities of trigonometry are algebraic deductions from these fundamental equations, all such identities also hold for the trigonometrical functions of a complex variable.

The hyperbolic functions of a complex variable are also defined in the same way as for real variables. The two fundamental ones, from which the others may be derived, are

$$\sinh z = \tfrac{1}{2}(e^z - e^{-z}), \; \cosh z = \tfrac{1}{2}(e^z + e^{-z}).$$

These two functions are clearly regular in any bounded domain.

The important relations

$$\sin iz = i \sinh z, \qquad \cos iz = \cosh z,$$
$$\sinh iz = i \sin z, \qquad \cosh iz = \cos z,$$

are easily proved and are of great usefulness for deducing properties of the hyperbolic functions from the corresponding properties of the trigonometrical functions.

If we write $z = x + iy$,

$$\sin z = \sin x \cosh y + i \cos x \sinh y,$$

and we see that $\sin z$ can only vanish if

$$\sin x \cosh y = 0, \; \cos x \sinh y = 0.$$

Now $\cosh y \geqslant 1$, and so the first equation implies that $\sin x$ is zero. Hence $x = n\pi$, $(n = 0, \pm 1, \pm 2, \ldots)$. The second then becomes $\sinh y = 0$, and this has only one root $y = 0$. Hence $\sin z$ vanishes if, and only if, $z = n\pi$, $(n = 0, \pm 1, \pm 2, \ldots)$. Similarly, we can show that $\cos z$ vanishes if, and only if, $z = (n + \tfrac{1}{2})\pi$.

IV. *The logarithmic function.*

When x is real and positive, the equation $e^u = x$ has one real solution $u = \log x$. If z is complex, however, but not zero, the corresponding equation $\exp w = z$ has an infinite number of solutions, each of which is called a *logarithm of z*. If $w = u + iv$ we have

$$e^u(\cos v + i \sin v) = z.$$

Hence we see that v is one of the values of arg z and $e^u = |z|$. Hence $u = \log |z|$. Every solution of exp $w = z$ is thus of the form

$$w = \log |z| + i \arg z.$$

Since arg z has an infinite number of values, there is an infinite number of logarithms of the complex number z, each pair differing by $2\pi i$. We write

$$\text{Log } z = \log |z| + i \arg z,$$

so that Log z is an infinitely many-valued function of z.

The **principal value** of Log z, which is obtained by giving arg z its principal value, will be denoted by log z, since it is identical with the ordinary logarithm when z is real and positive. We refer again to the logarithmic function in the next section, where many-valued functions are discussed in more detail.

V. *The general power* ζ^z.

So far we have only defined a^z when $a > 0$. If z and ζ denote any complex numbers we define the *principal value of the power* ζ^z, with $\zeta \neq 0$ as the only condition, to be the number uniquely determined by the equation

$$\zeta^z = e^{z \log \zeta},$$

where log ζ is the principal value of Log ζ. By choosing other values of Log ζ we obtain other values of the power which may be called its *subsidiary values*. All these are contained in the formula

$$\zeta^z = \exp\{z(\log \zeta + 2k\pi i)\}.$$

Hence ζ^z has an infinite number of values, in general, but one, and only one, principal value.

Example. i^i denotes the infinity of *real* numbers
$$\exp\{i(\log i + 2k\pi i)\} = \exp\{i(\tfrac{1}{2}\pi i + 2k\pi i)\}$$
$$= \exp(-\tfrac{1}{2}\pi - 2k\pi).$$
$\exp(-\tfrac{1}{2}\pi)$ is the *principal value* of the power i^i.

If $\zeta = 0$, R$z > 0$, we *define* ζ^z to be zero.

§ 9. Many-valued Functions

In the definition of a regular function given in § **7**, we note that a regular function must be *one-valued* (or *uniform*). Quite a number of elementary functions, such as z^a (a not an integer) log z, arc sin z are many-valued. To illustrate the idea of many-valuedness, let us consider the simple case of the relation $w^2 = z$. On putting $z = re^{i\theta}$, $w = Re^{i\phi}$, we get

$$R^2 e^{2i\phi} = re^{i\theta}.$$

For given r and $\theta(<2\pi)$, two obvious solutions are

$$w_1 = |\sqrt{r}| e^{\frac{1}{2}i\theta} \text{ and } w_2 = |\sqrt{r}| e^{i(\frac{1}{2}\theta + \pi)} = -|\sqrt{r}| e^{\frac{1}{2}i\theta},$$

and these are the only continuous solutions for fixed θ, since $|\sqrt{r}|$ and $-|\sqrt{r}|$ are the only continuous solutions of the real equation $x^2 = r$, $r > 0$.

In particular, for a *positive real z*, that is when $\theta = 0$, $w_1 = |\sqrt{r}|$ and $w_2 = -|\sqrt{r}|$, and both w_1 and w_2 are one-valued functions of z defined for all values of z.

If we follow the change in w_1 as θ varies from 0 to 2π, in other words, as the variable z describes a circle of radius r about the origin, w_1 varies continuously and we see that the final value of w_1 is $|\sqrt{r}| e^{\frac{1}{2} \cdot 2\pi i} = -|\sqrt{r}| = w_2$.

Hence the function w_1 is apparently discontinuous along the positive real axis, since the values just above and just below the real axis differ in sign and are not zero (except at the origin itself). If, however, z describes the circle round the origin a second time, the values of w_1 continue those of w_2 and at the end of the second circuit we have $w_2 = w_1$ along the positive real axis.

We thus see that the equation $w^2 = z$ has no continuous *one-valued* solution defined for the whole complex plane, but $w^2 = z$ defines a *two-valued* function of z. The two functions $w_1 = |\sqrt{r}| e^{\frac{1}{2}i\theta}$ and $w_2 = -|\sqrt{r}| e^{\frac{1}{2}i\theta}$ are called the two **branches** of the two-valued function $w^2 = z$. Each of these branches is a one-valued function in the z-plane if we make a narrow slit, extending from the origin to infinity along the positive real axis, and distinguish between the values of the function at points on the upper and lower edges of the cut. If $OA = x_0$, in fig. 2, the value of w_1 at $A(\theta = 0)$ is $|\sqrt{x_0}|$ and the value

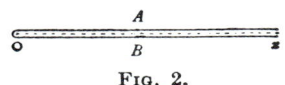

FIG. 2.

of w_1 at $B(\theta = 2\pi)$ is $-|\sqrt{x_0}|$. Since the cut effectively prevents the making of a complete circuit about the origin, if we start with a value of z belonging to the branch w_1, we can never change over to the branch w_2. Thus w_1 (and similarly w_2) is one-valued on the *cut-plane*.

There is an ingenious method of representing the two-valued function $w^2 = z$ as a one-valued function, by constructing what is known as a **Riemann surface**. This is equivalent to replacing the ordinary z-plane by two planes P_1 and P_2: we may think of P_1 as superposed on P_2. If we make a cut, as described above, in the two planes, we make the convention that the lower edge of the cut in P_1 shall be connected to the upper edge of the cut in P_2 and the lower edge of the cut in P_2 to the upper edge of the cut in P_1. Suppose that we start with a value z_0 of z, w_1^0 being the corresponding value of w_1 on the plane P_1, and let the point z describe a path, starting from z_0, in the counter-clockwise sense. When the moving point reaches the lower edge of the cut in P_1 it crosses to the upper edge of the cut in P_2, then describes another counter-clockwise circuit in P_2 until it reaches the lower edge of the cut in this plane. It then crosses again to the upper edge of the cut in P_1 and returns to its starting point with the same value w_1^0 with which it started. This corresponds precisely to the way in which we obtain the two different values of $\sqrt{z}$, and so $\sqrt{z}$ is *a one-valued function of position on the Riemann surface.*

There is no unique way of dividing up the function into branches, and we might have cut the plane along any line extending from the origin to infinity, but the point $z = 0$ is distinguished, for the function $w = \sqrt{z}$, from all other points, as we shall now see.

We observe that, if z describes a circle about any point a and the origin lies *outside* this circle, then arg z is not increased by 2π but returns to its initial value. Hence the values of w_1 and w_2 are exchanged only when z turns *about the origin*. For this reason the point $z = 0$ is called a **branch-point** of the function $w = \sqrt{z}$ and, as we have already stated, $w_1(z)$ and $w_2(z)$ are called its two branches.

Since turning about $z = \infty$ means, by definition, describing a large circle about the origin, the point $z = \infty$ is also a (conventional) branch-point for $w = \sqrt{z}$.

The relation $w^n = z$ defines an n-valued function of z, since $z^{\frac{1}{n}}$ has n, and only n, different values

$$r^{\frac{1}{n}} \operatorname{cis} \frac{\theta + 2s\pi}{n} \, , \, (s = 0, 1, 2, ..., n-1).$$

The point $z = 0$ is a branch-point, and the Riemann surface appropriate to this function consists of n sheets $P_1, P_2, ..., P_n$. Plainly $z = 1$ is a branch-point for $w = \sqrt{(z-1)}$ and the cut is made from $z = 1$ to $z = \infty$.

For $w = \operatorname{Log} z$, since $w = \log r + i(\theta + 2k\pi)$, every positive and negative integer k gives a branch, so $\operatorname{Log} z$ is an infinitely many-valued function of z. The Riemann surface consists of an infinity of superposed planes, each cut along the positive real axis, and each edge of each cut is joined to the opposite edge of the one below. The points $z = 0$ and $z = \infty$ are branch points.

For $w = \sqrt{\{(z-a)(z-b)\}}$ we make a cut on each plane along the straight line joining the points $z = a$ and $z = b$, and join the planes P_1 and P_2 cross-wise along the cut. In this case infinity is not a branch-point.

If $w = \sqrt{\{(z-a_1)(z-a_2)...(z-a_k)\}}$; when k is even, we make cuts joining pairs of points a_r, a_s; and when k is odd, one of these points must be joined to ∞, as in the case $k = 1$. The edges of the various cuts on the two planes are joined cross-wise as in the case $w = \sqrt{z}$.

Considerable ingenuity is required in constructing Riemann surfaces for more complicated functions, but it is beyond our scope to pursue this question further.

Note on notation. In what follows we shall frequently use w, z and ζ to denote complex numbers, and, whenever they are used, it will be understood, without further explanation, that

$$w \equiv u+iv,\ z \equiv x+iy,\ \zeta \equiv \xi+i\eta.$$

Other symbols, such as t, r, s, may occasionally be used to denote complex numbers, but we do not specify any special symbols to denote their real and imaginary parts.

EXAMPLES I

1. Prove that $|z_1-z_2|^2 + |z_1+z_2|^2 = 2|z_1|^2 + 2|z_2|^2$; and deduce that

$$|a+\sqrt{(a^2-\beta^2)}| + |a-\sqrt{(a^2-\beta^2)}| = |a+\beta| + |a-\beta|,$$

all the numbers concerned being complex.

2. Prove that the area of the triangle whose vertices are the points z_1, z_2, z_3 on the Argand diagram is

$$\Sigma\{(z_2-z_3)|z_1|^2/4iz_1\}.$$

Show also that the triangle is equilateral if

$$z_1^2+z_2^2+z_3^2 = z_1z_2+z_2z_3+z_3z_1.$$

3. Determine the regions of the Argand diagram defined by

$$|z^2-z|<1\ ;\ |z-a|+|z-b|\leqslant k\ (k>0)\ ;\ |z^2+az+b|<r^2.$$

In the last case, show that, if z_1, z_2 are the roots of $z^2+az+b=0$, we obtain two regions if $r<\frac{1}{2}|z_1-z_2|$.

4. A point $P(a+ib)$ lies on the line AB, where A is $z = \rho$ and B is $z = 2i\rho$. If Q is $-\rho^2/(a+ib)$, find the polar coordinates of P and Q referred to the origin O as pole and the real axis as initial line. Indicate the positions of P and Q in an Argand diagram. If P move along the line AB and C is the point $z = -\rho$, prove that the triangles OAP, OQC are similar, and that the locus of Q is a circle.

5. If
$$f(z) = \frac{x^3 y(y-ix)}{x^6+y^2} \ (z \neq 0), f(0) = 0,$$

prove that $\{f(z)-f(0)\}/z \to 0$ as $z \to 0$ along any radius vector, but not as $z \to 0$ *in any manner*.

6. Prove that the function $u+iv = f(z)$, where
$$f(z) = \frac{x^3(1+i)-y^3(1-i)}{x^2+y^2} \ (z \neq 0), f(0) = 0,$$

is continuous and that the Cauchy-Riemann equations are satisfied at the origin, yet $f'(0)$ does not exist.

7. Show that the function $f(z) = \sqrt{|xy|}$ is not regular at the origin, although the Cauchy-Riemann equations are satisfied at that point.

8. If $f(z)$ is a regular function of z, prove that
$$\left(\frac{\partial^2}{\partial x^2} + \frac{\partial^2}{\partial y^2}\right) |f(z)|^2 = 4 |f'(z)|^2.$$

9. If $w = f(z)$ is a regular function of z such that $f'(z) \neq 0$, prove that
$$\left(\frac{\partial^2}{\partial x^2} + \frac{\partial^2}{\partial y^2}\right) \log |f'(z)| = 0.$$

If $|f'(z)|$ is the product of a function of x and a function of y, show that
$$f'(z) = \exp(az^2+\beta z+\gamma)$$

where a is a real and β and γ are complex constants.

10. Prove that the functions

   (i) $u = x^3-3xy^2+3x^2-3y^2+1$,
   (ii) $u = \sin x \cosh y + 2 \cos x \sinh y + x^2 - y^2 + 4xy$,

both satisfy Laplace's equation, and determine the corresponding regular function $u+iv$ in each case.

11. If $w = \text{arc sin } z$, show that $w = n\pi \pm i\text{Log}\{iz + \sqrt{(1-z^2)}\}$ according as the integer n is even or odd, a cross-cut being made along the real axis from 1 to ∞, and from $-\infty$ to -1 to ensure the one-valuedness of the logarithm.

12. If $w = \sqrt{\{(1-z)(1+z^2)\}}$, A the point $(2, 0)$ and P a point in the first quadrant, prove that, if the value of w when $z = 0$ is 1, and z describes the curve OPA, the value of w at A is $-i\sqrt{5}$.

13. If $w = \sqrt{(2-2z+z^2)}$, and z describes a circle of centre $z = 1+i$ and radius $\sqrt{2}$ in the positive sense, determine the value of w (i) when z returns to O, (ii) when z crosses the axis of y, given that z starts from O with the value $+\sqrt{2}$ of w.

14. Prove that $\log(1+z)$ is regular in the z-plane, cut along the real axis from $-\infty$ to -1, and that this function can be expanded in a power series

$$z - \frac{z^2}{2} + \frac{z^3}{3} - \frac{z^4}{4} + \cdots,$$

convergent when $|z| < 1$.

15. Prove that the function

$$f(z) = 1 + \sum_{n=1}^{\infty} \frac{a(a-1)\dots(a \quad n+1)}{n!} z^n$$

is regular when $|z| < 1$ and that its derivative is $af(z)/(1+z)$. Hence deduce that $f(z) = (1+z)^a$.

16. (i) Prove that the exponential function e^z is a one-valued function of z.

(ii) Show that the values of $z = a^\zeta$, when plotted on the Argand diagram for z, are the vertices of an equiangular polygon inscribed in an equiangular spiral whose angle is independent of a.

17. If $0 < a_0 < a_1 < \dots < a_n$, prove that all the roots of the equation

$$a_0 z^n + a_1 z^{n-1} + \dots + a_n = 0$$

lie outside the circle $|z| = 1$.

18. Show that, if θ is real and $\sin \theta \sin \phi = 1$, then

$$\phi = (n+\tfrac{1}{2})\pi \pm i \log \cot \tfrac{1}{2}(n\pi + \theta),$$

where n is an integer, even or odd, according as $\sin \theta > 0$ or $\sin \theta < 0$. [If $\phi = a + i\beta$, we have $\sin a \cosh \beta = \text{cosec } \theta$, $\cos a \sinh \beta = 0$. Solve for a and β.]

CHAPTER II

CONFORMAL REPRESENTATION

§ 10. Isogonal and Conformal Transformations

The equations $u = u(x, y)$, $v = v(x, y)$ may be regarded as setting up a *correspondence* between a domain D of the (x, y)-plane and a domain D' of the (u, v)-plane. If the functions u and v are continuous, and possess continuous partial derivatives of the first order at each point of D, then any curve in D, which has a continuously turning tangent, corresponds to a curve in D' possessing the same property, but the correspondence between the two domains is not necessarily one-one.

For example, if $u = x^2$, $v = y^2$, the circular domain $x^2+y^2 \leqslant 1$ corresponds to the triangle formed by the lines $u = 0$, $v = 0$, $u+v = 1$, but there are four points of the circle corresponding to each point of the triangle.

If two curves in the domain D intersect at the point P, (x_0, y_0) at an angle θ, then, if the two corresponding curves in D' intersect at the point (u_0, v_0) corresponding to P *at the same angle θ*, the transformation is said to be **isogonal**. *If the sense of the rotation as well as the magnitude of the angle is preserved, the transformation is said to be* **conformal**.

Some writers do not distinguish between *isogonal* and *conformal*, but define conformality as the preservation of the magnitude of the angles without considering the sense.

If two domains correspond to each other by a given transformation $u = u(x, y)$, $v = v(x, y)$, then any figure in D may be said to be **mapped** on the corresponding figure in D' by means of the given transformation. We have already defined isogonal and conformal mapping, but it should be observed that, if one domain is mapped isogonally or conformally upon another, the correspondence between the domains is not necessarily *one-one*. If to each point of D there corresponds one, and only one, point of D', and conversely, the mapping of D on D', or of D' on D, is said to be **one–one or biuniform.**

Suppose that $w = f(z)$ is regular in a domain D of the z-plane, z_0 is an interior point of D, and C_1 and C_2 are two continuous curves passing through the point z_0, and let the tangents at this point make angles a_1, a_2 with the real axis. Our object is to discover what is the map of this figure on the w-plane. For a reason which will appear in a moment, *we suppose that $f'(z_0) \neq 0$.*

Let z_1 and z_2 be points on the curves C_1 and C_2 near to z_0 and at the same distance r from z_0, so that

$$z_1 - z_0 = re^{i\theta_1}, \; z_2 - z_0 = re^{i\theta_2},$$

then, as $r \to 0$, $\theta_1 \to a_1$ and $\theta_2 \to a_2$.

The point z_0 corresponds to a point w_0 in the w-plane and z_1 and z_2 correspond to points w_1 and w_2 which describe curves Γ_1 and Γ_2. Let

$$w_1 - w_0 = \rho_1 e^{i\phi_1}, \; w_2 - w_0 = \rho_2 e^{i\phi_2},$$

then, by the definition of a regular function,

$$\lim_{z_1 \to z_0} \frac{w_1 - w_0}{z_1 - z_0} = f'(z_0),$$

and, since the right-hand side is not zero, we may write it $Re^{i\lambda}$. We have

$$\lim \frac{\rho_1 e^{i\phi_1}}{re^{i\theta_1}} = Re^{i\lambda},$$

and so $\lim (\phi_1 - \theta_1) = \lambda$ or

$$\lim \phi_1 = a_1 + \lambda.$$

Thus we see that the curve Γ_1 has a definite tangent at w_0 making an angle $\alpha_1 + \lambda$ with the real axis.

Similarly, Γ_2 has a definite tangent at w_0 making an angle $\alpha_2 + \lambda$ with the real axis.

It follows that the curves Γ_1 and Γ_2 cut at the same angle as the curves C_1 and C_2. Further, the angle between the curves Γ_1, Γ_2 has the same sense as the angle between C_1, C_2. Thus the regular function $w = f(z)$, for which $f'(z_0) \neq 0$, determines a conformal transformation. Any small figure in one plane corresponds to an approximately similar figure in the other plane. To obtain one figure from the other we have to rotate it through the angle $\lambda = \arg\{f'(z_0)\}$ and subject it to the magnification

$$\lim \frac{\rho_1}{r} = R = |f'(z_0)|.$$

It is clear that the magnification is the same in all directions through the same point, but it varies from one point to another.

If ζ is a regular function of w and w is a regular function of z, then ζ is a regular function of z, and so, if a region of the z-plane is represented conformally on a region of the w-plane and this in its turn on a region of the ζ-plane, the transformation from the z-plane to the ζ-plane will be conformal.

There exist transformations in which the magnitude of the angles is conserved but their sign is changed. For example, consider the transformation

$$w = x - iy \, ;$$

this replaces every point by its reflection in the real axis and so, while angles are conserved, their signs are changed. This is true generally for every transformation of the form

$$w = f(\bar{z}) \qquad \cdot \qquad \cdot \qquad \cdot \qquad \cdot \qquad (1)$$

where $f(z)$ is regular; for it is a combination of two transformations

$$\text{(i) } \zeta = \bar{z}, \text{ (ii) } w = f(\zeta).$$

In (i) angles are conserved but their signs are changed, and in (ii) angles and signs are conserved. Hence in the given transformation, angles are conserved and their signs changed. *Thus* (1) *gives a transformation which is isogonal but not conformal.*

We have seen that every regular function $w = f(z)$, defined in a domain in which $f'(z)$ is not zero, maps the domain in the z-plane *conformally* on the corresponding domain in the w-plane. Let us now consider the problem from the converse point of view. *Given a pair of differentiable relations of the type*

$$u = u(x, y), \ v = v(x, y) \qquad . \qquad . \quad (2)$$

defining a transformation from (x, y)-*space to* (u, v)-*space does there correspond a regular function* $w = f(z)$?

Let $d\sigma$ and ds be elements of length in the (u, v)-plane and (x, y)-plane respectively. Then $d\sigma^2 = du^2 + dv^2$, $ds^2 = dx^2 + dy^2$ and so, since

$$du = \frac{\partial u}{\partial x} dx + \frac{\partial u}{\partial y} dy, \ dv = \frac{\partial v}{\partial x} dx + \frac{\partial v}{\partial y} dy;$$

$$d\sigma^2 = E dx^2 + 2F dx dy + G dy^2,$$

where

$$E = \left(\frac{\partial u}{\partial x}\right)^2 + \left(\frac{\partial v}{\partial x}\right)^2, \ F = \frac{\partial u}{\partial x} \frac{\partial u}{\partial y} + \frac{\partial v}{\partial x} \frac{\partial v}{\partial y},$$

$$G = \left(\frac{\partial u}{\partial y}\right)^2 + \left(\frac{\partial v}{\partial y}\right)^2.$$

Then the ratio $d\sigma : ds$ is independent of direction if

$$\frac{E}{1} = \frac{F}{0} = \frac{G}{1}.$$

On writing h^2 for E (or G), where h depends only on x and y and is not zero, the conditions for an isogonal transformation are

$$\left(\frac{\partial u}{\partial x}\right)^2 + \left(\frac{\partial v}{\partial x}\right)^2 = \left(\frac{\partial u}{\partial y}\right)^2 + \left(\frac{\partial v}{\partial y}\right)^2 = h^2,$$

$$\frac{\partial u}{\partial x}\frac{\partial u}{\partial y} + \frac{\partial v}{\partial x}\frac{\partial v}{\partial y} = 0.$$

The first two equations are satisfied by writing $u_x = h\cos\alpha$, $v_x = h\sin\alpha$, $u_y = h\cos\beta$, $v_y = h\sin\beta$, and the third is plainly satisfied if

$$\alpha - \beta = \pm\tfrac{1}{2}\pi.$$

Hence the correspondence is isogonal if either

(a) $u_x = v_y$, $v_x = -u_y$ or (b) $u_x = -v_y$, $v_x = u_y$.

Equations (a) are the Cauchy-Riemann equations and express that $u+iv = f(x+iy)$ where $f(z)$ is a regular function of z. Equations (b) reduce to (a) by writing $-v$ for v, that is, by taking the image figure found by reflection in the real axis of the w-plane. Hence (b) corresponds to an isogonal, but not conformal transformation, and so it follows that the only *conformal* transformations of a domain in the z-plane into a domain of the w-plane are of the form $w = f(z)$ where $f(z)$ is a regular function of z.

The case $f'(z) = 0$.

We laid down above the condition that $f'(z_0) \neq 0$. Suppose now that $f'(z)$ has a zero * of order n at the point z_0, then, in the neighbourhood of this point,

$$f(z) = f(z_0) + a(z-z_0)^{n+1} + \ldots$$

where $a \neq 0$. Hence

$$w_1 - w_0 = a(z_1 - z_0)^{n+1} + \ldots$$

or $$\rho_1 e^{i\phi_1} = |a| r^{n+1} e^{i\{\lambda + (n+1)\theta_1\}} + \ldots$$

* See § 36.

where $\lambda = \arg a$. Hence

$$\lim \phi_1 = \lim\{\lambda + (n+1)\theta_1\}$$
$$= \lambda + (n+1)a_1.$$

Similarly, $\lim \phi_2 = \lambda + (n+1)a_2$.

Thus the curves Γ_1, Γ_2 still have definite tangents at w_0, but the angle between the tangents is

$$\lim (\phi_2 - \phi_1) = (n+1)(a_2 - a_1).$$

Also the linear magnification, $\lim (\rho_1/r)$, is zero. Hence the conformal property does not hold at such a point.

For example, consider $w = z^2$. In this $\arg w = 2 \arg z$ and the angle between the line joining the origin to the point w_0 and the positive real axis is double the angle between the line joining the origin to the corresponding point z_0 and the positive real axis in the z-plane. Corresponding angles at the origins are not equal because, at $z = 0$, $dw/dz = 0$.

Points at which $dw/dz = 0$ or ∞ will be called **critical points** of the transformation defined by $w = f(z)$. These points play an important part in the transformations.

§ 11. Harmonic Functions

Solutions of Laplace's equation, $\nabla^2 V = 0$, are called **harmonic** functions ; and, in applications to mathematical physics, an important problem to be solved is that of finding a function which is harmonic in a given domain and takes given values on the boundary. This is known as Dirichlet's problem. In the three-dimensional case, if we make a transformation from (x, y, z)-space to (ξ, η, ζ)-space, it will, in general, alter Dirichlet's problem. If $V(x, y, z)$ is harmonic, and we make the transformation

$$x = \phi_1(\xi, \eta, \zeta),\ y = \phi_2(\xi, \eta, \zeta),\ z = \phi_3(\xi, \eta, \zeta),$$

the function $V_1(\xi, \eta, \zeta)$, into which $V(x, y, z)$ is transformed is not, in general, harmonic in (ξ, η, ζ)-space. In two-dimensional problems, however, if the transformation is conformal, Laplace's equation in (x, y)-space corresponds to Laplace's equation in (u, v)-space, and the problem to be solved in (u, v)-space is still Dirichlet's problem. To prove this, consider a transformation

$$u = u(x, y), \quad v = v(x, y) . \qquad \bullet \qquad \bullet \qquad (1)$$

where $w = u + iv$ is a regular function of $z = x + iy$, say $w = f(z)$. Let D be a domain of the (x, y)-plane throughout which $f'(z) \neq 0$, and let $\varDelta$ be the domain of the (u, v)-plane which corresponds to D by means of the given transformation.

If x and y are the independent variables, and V any twice-differentiable function of x and y, we have *

$$d^2V = \frac{\partial^2 V}{\partial u^2} du^2 + 2 \frac{\partial^2 V}{\partial u \partial v} du dv + \frac{\partial^2 V}{\partial v^2} dv^2 + \frac{\partial V}{\partial u} d^2u + \frac{\partial V}{\partial v} d^2v, \quad (2)$$

and

$$du = \frac{\partial u}{\partial x} dx + \frac{\partial u}{\partial y} dy, \quad dv = \frac{\partial v}{\partial x} dx + \frac{\partial v}{\partial y} dy, \quad . \quad (3)$$

$$d^2u = \frac{\partial^2 u}{\partial x^2} dx^2 + 2 \frac{\partial^2 u}{dxdy} dxdy + \frac{\partial^2 u}{dy^2} dy^2 ; \quad . \qquad \bullet \quad (4)$$

with a similar expression for d^2v.

On substituting for du, dv, d^2u, d^2v the expressions (3) and (4), the expression (2) for d^2V becomes a quadratic expression in the differentials of the independent variables dx and dy and, on selecting the coefficients of dx^2 and dy^2, we get

$$\frac{\partial^2 V}{\partial x^2} = \frac{\partial^2 V}{\partial u^2} \left(\frac{\partial u}{\partial x}\right)^2 + 2 \frac{\partial^2 V}{\partial u \partial v} \frac{\partial u}{\partial x} \frac{\partial v}{\partial x} + \frac{\partial^2 V}{\partial v^2} \left(\frac{\partial v}{\partial x}\right)^2 + \frac{\partial V}{\partial u} \frac{\partial^2 u}{\partial x^2} + \frac{\partial V}{\partial v} \frac{\partial^2 v}{\partial x^2},$$

$$\frac{\partial^2 V}{\partial y^2} = \frac{\partial^2 V}{\partial u^2} \left(\frac{\partial u}{\partial y}\right)^2 + 2 \frac{\partial^2 V}{\partial u \partial v} \frac{\partial u}{\partial y} \frac{\partial v}{\partial y} + \frac{\partial^2 V}{\partial v^2} \left(\frac{\partial v}{\partial y}\right)^2 + \frac{\partial V}{\partial u} \frac{\partial^2 u}{\partial y^2} + \frac{\partial V}{\partial v} \frac{\partial^2 v}{\partial y^2}.$$

But we have already seen that u and v satisfy the Cauchy-Riemann equations

$$\frac{\partial u}{\partial x} = \frac{\partial v}{\partial y}, \quad \frac{\partial v}{\partial x} = -\frac{\partial u}{\partial y},$$

* See P.A., p. 233.

and also Laplace's equation.   On addition we therefore get

$$\frac{\partial^2 V}{\partial x^2} + \frac{\partial^2 V}{\partial y^2} = \left(\frac{\partial^2 V}{\partial u^2} + \frac{\partial^2 V}{\partial v^2}\right)\left\{\left(\frac{\partial u}{\partial x}\right)^2 + \left(\frac{\partial u}{\partial y}\right)^2\right\}.$$

Now $f'(z) = u_x + iv_x$, and so the last bracket is equal to $|f'(z)|^2$, the square of the linear magnification of the transformation.   Since this is not zero, it follows that if

$$\frac{\partial^2 V}{\partial x^2} + \frac{\partial^2 V}{\partial y^2} = 0,$$

then

$$\frac{\partial^2 V}{\partial u^2} + \frac{\partial^2 V}{\partial v^2} = 0.$$

§ 12. Superficial Magnification

We have already seen that the linear magnification at any point in a conformal transformation $w = f(z)$ is $|f'(z)|$, it being supposed that $f'(z) \neq 0$.   We now prove that the superficial magnification is $|f'(z)|^2$.   If $\varDelta$ be the closed domain of the w-plane which corresponds to a closed domain D of the z-plane, the area A of $\varDelta$ is given by

$$A = \iint_\varDelta du\,dv = \iint_D \left|\frac{\partial(u,\,v)}{\partial(x,\,y)}\right| dx\,dy,$$

by the well-known theorem for change of variables in a double integral.*   Now

$$\frac{\partial(u,\,v)}{\partial(x,\,y)} = \frac{\partial u}{\partial x}\frac{\partial v}{\partial y} - \frac{\partial v}{\partial x}\frac{\partial u}{\partial y} = \left(\frac{\partial u}{\partial x}\right)^2 + \left(\frac{\partial u}{\partial y}\right)^2$$

by the Cauchy-Riemann equations : but, as we have just seen,

$$\left(\frac{\partial u}{\partial x}\right)^2 + \left(\frac{\partial u}{\partial y}\right)^2 = \left|\frac{\partial u}{\partial x} + i\,\frac{\partial v}{\partial x}\right|^2 = |f'(z)|^2$$

and so

$$A = \iint_D |f'(z)|^2 dx\,dy.$$

This proves the theorem.

* P.A., p. 302 ; or Gillespie, *Integration*, p. 40. (Hereafter cited as G.I.)

§ 13. The Bilinear Transformation

We have seen that a regular function $w = f(z)$, for which $f'(z_0) \neq 0$, gives a continuous one-one representation of a certain neighbourhood of the point z_0 of the z-plane on a neighbourhood of a point w_0 of the w-plane, and that this representation is conformal. It may be expressed in another way by saying that by such a transformation *infinitely small* circles of the z-plane correspond to infinitely small circles of the w-plane. There are, however, non-trivial conformal transformations for which this is true of *finite* circles : these transformations will now be investigated.

Let A, B, C denote three complex constants, $\bar{A}$, $\bar{B}$, $\bar{C}$ their conjugates and z, $\bar{z}$ a complex variable and its conjugate ; then the equation

$$(A+\bar{A})z\bar{z}+Bz+\bar{B}\bar{z}+C+\bar{C} = 0 \qquad . \qquad . \quad (1)$$

represents a real circle or a straight line, provided that

$$B\bar{B} > (A+\bar{A})(C+\bar{C}). \qquad . \qquad . \qquad . \quad (2)$$

For, if we write $A = a+ia'$, $B = b+ib'$, $C = c+ic'$, $z = x+iy$, (1) becomes

$$a(x^2+y^2)+bx-b'y+c = 0$$

which is the equation of a circle. It reduces to a straight line if $a = \frac{1}{2}(A+\bar{A}) = 0$. If r be the radius of this circle,

$$r^2 = \frac{b^2}{4a^2} + \frac{b'^2}{4a^2} - \frac{c}{a},$$

and the circle is real provided that

$$b^2+b'^2 > 4ac,$$

which is the same as condition (2).

Conversely, every real circle or straight line can, by suitable choice of the constants, be represented by an equation of the form (1) satisfying the condition (2).

The transformation

$$w = \frac{az+\beta}{\gamma z+\delta} \qquad \cdot \qquad \cdot \qquad \cdot \qquad \cdot \qquad (3)$$

where a, β, γ, δ are complex constants is called a **bilinear transformation**. It is the most general type of transformation for which one and only one value of z corresponds to each value of w, and conversely. Since the bilinear transformation (3) was first studied by Möbius (1790-1868) we shall, following Carathéodory,* call it also a **Möbius' transformation**.

The expression $a\delta-\beta\gamma$, called the **determinant** of the transformation, must not vanish. If $a\delta-\beta\gamma = 0$, the right-hand side of (3) is either a constant or meaningless. For convenience it is sometimes agreed to arrange that $a\delta-\beta\gamma = 1$. The determinant in the general case can always be made to have the value unity if the numerator and denominator of the fraction on the right-hand side of (3) be divided by $\pm\sqrt{(a\delta-\beta\gamma)}$.

If we write

$$w = (a/\gamma)+(\beta\gamma-a\delta)(\omega/\gamma), \; \omega = 1/\zeta, \; \zeta = \gamma z+\delta \quad . \quad (4)$$

it is easily seen that (3) is equivalent to the succession of transformations (4).

Now $w = z+a$ corresponds to a *translation*, since the figure in the w-plane is merely the same figure as in the z-plane with a different origin.

Consider next $w = \rho z$, where ρ is real. The two figures in the z-plane and the w-plane are similar and similarly situated about their respective origins, but the scale of the w-figure is ρ times that of the z-figure. Such a transformation is a *magnification*.

In the third place we consider $w = ze^{i\theta}$. Clearly $|w| = |z|$ and one value of $\arg w$ is $\theta+\arg z$, and so the w-figure is the z-figure turned about the origin through

* *Conformal Representation* (Cambridge, 1932).

an angle θ in the positive sense. Such a transformation is a *rotation*.

Finally, consider $w = 1/z$. If $|z| = r$ and $\arg z = \theta$, then $|w| = 1/r$ and $\arg w = -\theta$. Hence, to pass from the w-figure to the z-figure we *invert the former with respect to the origin of the w-plane, with unit radius of inversion, and then construct the image figure in the real axis of the w-plane*.

The sequence of transformations (4) consists of a combination of the ones just considered; of these, the only one which affects the shape of the figures is inversion. Since the inverse of a circle is a circle or a straight line (circle with infinite radius), it follows that a Möbius' transformation transforms circles into circles.

The transformation inverse to (3) is also a Möbius' transformation

$$z = \frac{-\delta w + \beta}{\gamma w - a}, \quad (-\delta)(-a) - \beta\gamma \neq 0 . \qquad . \quad (5)$$

Further, if we perform first the transformation (3), then a second Möbius' transformation

$$\zeta = \frac{a'w + \beta'}{\gamma'w + \delta'}, \quad a'\delta' - \beta'\gamma' \neq 0,$$

the result is a third Möbius' transformation

$$\zeta = \frac{Az + B}{\Gamma z + \Delta}$$

where $A\Delta - B\Gamma = (a\delta - \beta\gamma)(a'\delta' - \beta'\gamma') \neq 0$.

Since the right-hand side of (3) is a regular function of z, except when $z = -\delta/\gamma$, Möbius' transformations are conformal.

If we write the equation of a circle (1) in the form

$$Az\bar{z} + Bz + \bar{B}\bar{z} + C = 0 . \qquad . \qquad . \quad (6)$$

where $\mathcal{A}$ and $\mathcal{C}$ are real, and then substitute

$$z = \frac{-\delta w + \beta}{\gamma w - \alpha} \ , \ \ \bar{z} = \frac{-\delta \bar{w} + \bar{\beta}}{\bar{\gamma}\bar{w} - \bar{\alpha}}$$

in (6) we get an expression of the form

$$Dw\bar{w} + Ew + \bar{E}\bar{w} + F = 0 \quad . \qquad \bullet \qquad \bullet \qquad (7)$$

where

$$D \equiv \mathcal{A}\delta\bar{\delta} - B\bar{\gamma}\delta - \bar{B}\gamma\delta + \mathcal{C}\gamma\bar{\gamma}$$

and

$$F \equiv \mathcal{A}\beta\bar{\beta} - B\bar{\alpha}\beta - \bar{B}\alpha\bar{\beta} + \mathcal{C}\alpha\bar{\alpha}$$

are both real, and it is easy to verify that the coefficients of w and $\bar{w}$ are conjugate complex numbers. Hence (7) represents a circle in the w-plane, since it is of the same form as (1).

§ 14. Geometrical Inversion

From what we have just seen it would be natural to expect that there would be an intimate relation between Möbius' transformations and geometrical inversion.

Let S be a circle of centre K and radius r in the z-plane. Then two points P and P_1, collinear with K, such that $KP \cdot KP_1 = r^2$, are called **inverse points** with respect to the circle S, and it is known from geometry that any circle passing through P and P_1 is orthogonal to S. In the case of a straight line s, P and P_1 are inverse points with respect to s, if P_1 is the image of P in s. If P, P_1, and K are the points z, z_1, and k we have

$$|(z_1 - k)(z - k)| = r^2, \ \ \arg (z_1 - k) = \arg (z - k), \quad . \quad (8)$$

the second equation expressing the collinearity of the points K, P, P_1. The two equations (8) are satisfied, if, and only if,

$$(z_1 - k)(\bar{z} - \bar{k}) = r^2. \quad . \qquad \bullet \qquad \bullet \qquad (9)$$

If S is the circle

$$\mathcal{A}z\bar{z} + Bz + \bar{B}\bar{z} + \mathcal{C} = 0, \quad . \qquad \bullet \qquad \bullet \qquad (10)$$

which may be written

$$\left(z + \frac{\bar{B}}{A}\right)\left(\bar{z} + \frac{B}{A}\right) = \frac{B\bar{B} - A\mathcal{C}}{A^2},$$

we see that (10) is a circle with centre $-\bar{B}/A$ and radius

$$\sqrt{\left(\frac{B\bar{B} - A\mathcal{C}}{A^2}\right)}.$$

Hence equation (9) becomes

$$\left(z_1 + \frac{\bar{B}}{A}\right)\left(\bar{z} + \frac{B}{A}\right) = \frac{B\bar{B} - A\mathcal{C}}{A^2},$$

which on simplification is

$$A z_1 \bar{z} + B z_1 + \bar{B}\bar{z} + \mathcal{C} = 0. \qquad \qquad (11)$$

We thus get the relation between z and its inverse z_1 from the equation of S by substituting z_1 for z and leaving $\bar{z}$ unchanged. On solving (11), the transformation is

$$z_1 = \frac{-\bar{B}\bar{z} - \mathcal{C}}{A\bar{z} + B} \qquad \qquad (12)$$

We now prove the theorem :

The bilinear transformation transforms two points which are inverse with respect to a circle into two points which are inverse with respect to the transformed circle.

If z and z_1 are inverse with respect to the circle (10) then (11) is satisfied. Make the transformation (3) and let w and w_1 be the transformed points. We have

$$z_1 = \frac{-\delta w_1 + \beta}{\gamma w_1 - \alpha}, \quad \bar{z} = \frac{-\delta \bar{w} + \bar{\beta}}{\bar{\gamma}\bar{w} - \bar{\alpha}}$$

and, on substituting these values in (11) we get an expression

$$D w_1 \bar{w} + E w_1 + \bar{E}\bar{w} + F = 0$$

where the coefficients D, E, $\bar{E}$, F are the same as those of (7) ; in fact we get (7) with w replaced by w_1. But this is the condition that w and w_1 are inverse points with respect to the transformed circle (7). The theorem is therefore proved.

The inversion (12) can be written as a succession of two transformations

$$w = \bar{z}, \quad z_1 = \frac{-\bar{B}w - \bar{C}}{Aw + B}.$$

The first is a reflection in the real axis and the second is a Möbius' transformation. The first preserves the angles but reverses their signs; the second is conformal. Hence inversion is an isogonal, but not conformal, transformation.

Since inversion is a one-one isogonal, but not conformal, transformation, it is clear that the result of two, or of an even number of inversions, is a one-one conformal transformation, since both the magnitude and sign of the angles is preserved. In other words, the successive performance of an even number of inversions is equivalent to a Möbius' transformation.

§ 15. The Critical Points

If the z-plane is closed by the addition of the point $z = \infty$, then (3) and (5) show that every Möbius' transformation is a one-one transformation of the closed z-plane into itself.

If $\gamma \neq 0$ the point $w = a/\gamma$ corresponds to $z = \infty$ and $w = \infty$ to $z = -\delta/\gamma$; but, if $\gamma = 0$, the points $z = \infty$, $w = \infty$ correspond to each other. Since, from (4),

$$\frac{dw}{dz} = \frac{a\delta - \beta\gamma}{(\gamma z + \delta)^2},$$

the only critical points of the transformation are $z = \infty$ and $z = -\delta/\gamma$.

These two critical points cease to be exceptional if we extend the definition of conformal representation in the following manner. A function $w = f(z)$ is said to transform the neighbourhood of a point z_0 conformally into a neighbourhood of $w = \infty$, if the function $t = 1/f(z)$ transforms the neighbourhood of z_0 conformally into a neighbourhood of $t = 0$. Also $w = f(z)$ is said to transform the neighbourhood of $z = \infty$ conformally into a neighbourhood of w_0 if $w = \phi(\zeta) = f(1/\zeta)$ transforms the neighbourhood of $\zeta = 0$ conformally into a neighbourhood of w_0. In this definition w_0 may have the value ∞.

With these extensions of the definitions we may now say that *every Möbius' transformation gives a one-one conformal representation of the whole closed z-plane on the whole closed w-plane*. In other words, the mapping is *biuniform* for the complete planes of w and z.

§ 16. Coaxal Circles

Let a, b, z be the affixes of the three points A, B, P of the z-plane. Then

$$\arg \frac{z-b}{z-a} = A\hat{P}B,$$

if the principal value of the argument be chosen. Let A and B be fixed and P a variable point.

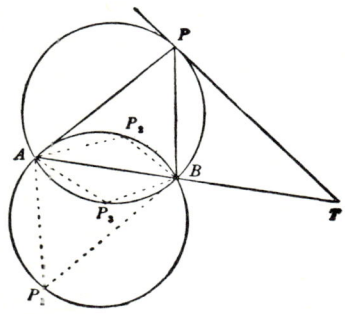

Fig. 3.

If the two circles in fig. 3 are equal, z_1, z_2, z_3 are the affixes of the points P_1, P_2, P_3 and $APB = \theta$, we see that

$$\arg \frac{z_2-b}{z_2-a} = \pi - \theta \,, \ \arg \frac{z_1-b}{z_1-a} = -\theta \,, \ \arg \frac{z_3-b}{z_3-a} = -\pi + \theta.$$

The locus defined by the equation

$$\arg \frac{z-b}{z-a} = \theta, \qquad \bullet \qquad \bullet \qquad \bullet \qquad (1)$$

when θ is a constant, is the arc APB. By writing $-\theta$, $\pi-\theta$, $-\pi+\theta$ for θ we obtain the arcs AP_1B, AP_2B, AP_3B respectively. The system of equations obtained by varying θ from $-\pi$ to π represents the system of circles which can be drawn through the points A, B. It should be observed that each circle must be divided into two parts, to each of which correspond different values of θ.

Let T be the point at which the tangent to the circle APB at P meets AB. Then the triangles TPA, TBP are similar and

$$\frac{AP}{PB} = \frac{PT}{BT} = \frac{TA}{TP} = k.$$

Hence $TA/TB = k^2$ and so T is a fixed point for all positions of P which satisfy

$$\left| \frac{z-a}{z-b} \right| = k, \qquad \bullet \qquad \bullet \qquad \bullet \qquad (2)$$

where k is a constant. Also $TP^2 = TA \cdot TB$ and so is constant. Hence the locus of P is a circle whose centre is T.

The system of equations obtained by varying k represents a system of circles. The system given by (1) is a system of coaxal circles of the common point kind, and that given by (2) a system of the limiting point kind, with A and B as the limiting points of the system. If $k \to \infty$ or if $k \to 0$ then the circle becomes a point circle at A or B. All the circles of one system intersect all the circles of the other system orthogonally.

The above important result is of frequent application in problems involving bilinear transformations. It may be used to prove that the bilinear transformation transforms circles into circles.

Suppose that the circle in the w-plane is

$$\left|\frac{w-\lambda}{w-\mu}\right| = k.$$

If we substitute for w in terms of z from the bilinear relation

$$w = \frac{az+\beta}{\gamma z+\delta}, \quad . \quad . \quad . \quad . \quad (3)$$

we obtain

$$\left|\frac{z-\lambda'}{z-\mu'}\right| = k',$$

where

$$\lambda' = -\frac{\beta-\lambda\delta}{a-\lambda\gamma}, \; \mu' = -\frac{\beta-\mu\delta}{a-\mu\gamma}, \; k' = \left|\frac{a-\mu\gamma}{a-\lambda\gamma}\right| k \,,$$

and so the locus in the z-plane is also a circle.

We may write (3) in the form

$$w = \frac{a}{\gamma} \frac{z+\beta/a}{z+\delta/\gamma},$$

and since

$$\left|\frac{z+\beta/a}{z+\delta/\gamma}\right| = k$$

represents a circle in the z-plane, the circle in the w-plane corresponding to it is plainly

$$|w| = k|a/\gamma|.$$

By taking special values of a, β, γ, δ and k the boundaries in the w-plane corresponding to given boundaries in the z-plane are easily determined.

For example, let $\beta/a = -i$, $\delta/\gamma = i$ and $k = 1$: since the locus $|(z-i)/(z+i)| = 1$ is plainly the real axis in the z-plane, this axis corresponds to the circle $|w| = |a/\gamma|$ in the w-plane: this will be the *unit* circle if, in addition, $|a| = |\gamma|$.

§ 17. Invariance of the Cross–Ratio

Let z_1, z_2, z_3, z_4 be any four points of the z-plane and let w_1, w_2, w_3, w_4 be the points which correspond to them by the Möbius' transformation

$$w = \frac{\alpha z + \beta}{\gamma z + \delta}. \qquad . \qquad . \qquad . \qquad (1)$$

If we suppose that all the numbers z_r, w_r are finite, we have

$$w_r - w_s = \frac{\alpha z_r + \beta}{\gamma z_r + \delta} - \frac{\alpha z_s + \beta}{\gamma z_s + \delta} = \frac{\alpha \delta - \beta \gamma}{(\gamma z_r + \delta)(\gamma z_s + \delta)} (z_r - z_s),$$

and hence it follows that

$$\frac{(w_1 - w_4)(w_3 - w_2)}{(w_1 - w_2)(w_3 - w_4)} = \frac{(z_1 - z_4)(z_3 - z_2)}{(z_1 - z_2)(z_3 - z_4)} \qquad . \qquad (2)$$

The right-hand side of (2) is the **cross–ratio** of the four points z_1, z_2, z_3, z_4, and so we have the result that the cross-ratio is invariant for the transformation (1).

If equation (2) be suitably modified, it is still true if any one of the numbers z_r or one of the numbers w_r is infinite. For example, let $z_2 = \infty$ and $w_1 = \infty$, then

$$\frac{w_3 - w_2}{w_3 - w_4} = \frac{z_1 - z_4}{z_3 - z_4} \qquad . \qquad . \qquad . \qquad (3)$$

Now suppose that z_r, w_r ($r = 1$, 2, 3) be two sets each containing three unequal complex numbers. Suppose first that these six numbers are all finite. Then the equation

$$\frac{(w_1 - w)(w_3 - w_2)}{(w_1 - w_2)(w_3 - w)} = \frac{(z_1 - z)(z_3 - z_2)}{(z_1 - z_2)(z_3 - z)}. \qquad . \qquad (4)$$

when solved for w leads to a Möbius' transformation which transforms each point z_r into the corresponding point w_r. The determinant of the transformation has the value

$$\alpha \delta - \beta \gamma = (w_1 - w_2)(w_1 - w_3)(w_2 - w_3)(z_1 - z_2)(z_1 - z_3)(z_2 - z_3) \neq 0.$$

It is also clear that (4) is the *only* Möbius' transformation which does so. The result still holds, if (4) be suitably modified, when one of the numbers z_r or w_r is infinite.

The equation (4) above may be used to find the particular transformations which transform one given circle into another given circle or straight line. A circle is uniquely determined by three points on its circumference and so we have only to give special values to each of the three sets z_r, $w_r (r = 1, 2, 3)$ and substitute them in (4).

Example. Let $z_1 = 1$, $z_2 = i$, $z_3 = -1$ and $w_1 = 0$, $w_2 = 1$, $w_3 = \infty$ then we get, after substitution in (4),

$$w = i\,\frac{1-z}{1+z}, \quad . \quad . \quad . \quad . \quad (5)$$

which transforms the circle $|z| = 1$ into the real axis of the w-plane and the *interior* of the circle $|z| < 1$ into the *upper* half of the w-plane.

The easiest way to prove this is as follows. Equation (5) is equivalent to

$$z = -\,\frac{w-i}{w+i}.$$

The boundary $|z| = 1$ corresponds to $|w-i| = |w+i|$, which is the real axis of the w-plane, since it is the locus of points equidistant from $w = \pm i$.

Since the centre $z = 0$ of the circle corresponds to the point $w = i$, in the upper half of the w-plane, the *interior* of the circle $|z| = 1$ corresponds to the *upper* half of the w-plane.

Similarly, since $w = -i$ corresponds to $z = \infty$, the *outside* of the circle $|z| = 1$ corresponds to the *lower* half of the w-plane.

It may be observed that although this use of the invariance of the cross-ratio will always determine the Möbius' transformation which transforms any given circle into any other given circle (or straight line), it is not necessarily the easiest way of doing so.

Thus, in the previous example, since $z = 1$ and -1 correspond to $w = 0$ and ∞, the transformation must take the form

$$w = \frac{k(z-1)}{z+1}.$$

Since $z = i$ corresponds to $w = 1$, we can determine k : thus

$$1 = \frac{k(i-1)}{i+1},$$

from which it readily follows that $k = -i$, and we obtain (5) above.

§ 18. Some special Möbius' Transformations

I. Let us consider the problem of finding *all the Möbius' transformations which transform the half-plane* $\mathbf{I}(z) \geqslant 0$ *into the unit circle* $|w| \leqslant 1$.

We observe first that to points z, $\bar{z}$ symmetrical with respect to the real axis correspond points w, $1/w$ inverse with respect to the unit circle in the w-plane. (See § **14**.) In particular, the origin and the point at infinity in the w-plane correspond to conjugate values of z. Let the required transformation be

$$w = \frac{az+\beta}{\gamma z+\delta}.$$

Clearly $\gamma \neq 0$, or the points at infinity in the two planes would correspond. Since $w = 0$, $w = \infty$ correspond to $z = -\beta/a$, $z = -\delta/\gamma$ we may write $-\beta/a = a$, $-\delta/\gamma = \bar{a}$ and

$$w = \frac{a}{\gamma} \frac{z-a}{z-\bar{a}}.$$

The point $z = 0$ must correspond to a point on the circle $|w| = 1$, so that

$$\left| \frac{a}{\gamma} \cdot \frac{-a}{-\bar{a}} \right| = \left| \frac{a}{\gamma} \right| = 1 \text{ ;}$$

hence we may write $a = \gamma e^{i\theta}$, where θ is real, and obtain

$$w = \frac{z-a}{z-\bar{a}}\, e^{i\theta} \qquad . \qquad . \qquad . \quad (1)$$

Since $z = a$ gives $w = 0$, a must be a point of the *upper* half-plane, in other words, $Ia > 0$. With this condition, (1) is the transformation required.

II. *To find all the Möbius' transformations which transform the unit circle $|z| \leqslant 1$ into the unit circle $|w| \leqslant 1$.*

Let $$w = \frac{az+\beta}{\gamma z+\delta}.$$

In this case $w = 0$ and $w = \infty$ must correspond to inverse points $z = a$, $z = 1/\bar{a}$, where $|a| < 1$. Hence $-\beta/a = a$, $-\delta/\gamma = 1/\bar{a}$, and so

$$w = \frac{a}{\gamma}\, \frac{z-a}{z-1/\bar{a}} = \frac{a\bar{a}}{\gamma}\, \frac{z-a}{\bar{a}z-1}.$$

The point $z = 1$ corresponds to a point on $|w| = 1$, and so

$$\left| \frac{a\bar{a}}{\gamma}\, \frac{1-a}{\bar{a}-1} \right| = \left| \frac{a\bar{a}}{\gamma} \right| = 1.$$

It follows that, if θ is real, $a\bar{a} = \gamma e^{i\theta}$ and so

$$w = \frac{z-a}{\bar{a}z-1}\, e^{i\theta}.$$

This is the desired transformation ; for, if $z = e^{i\psi}$, $a = \mathfrak{h}e^{i\lambda}$, then

$$|w| = \left| \frac{e^{i\psi}-be^{i\lambda}}{be^{i(\psi-\lambda)}-1} \right| = 1.$$

If $z = re^{i\psi}$, where $r < 1$, then

$$|z-a|^2-|\bar{a}z-1|^2$$
$$= r^2-2rb\cos(\psi-\lambda)+b^2-\{b^2r^2-2rb\cos(\psi-\lambda)+1\}$$
$$= (r^2-1)(1-b^2) < 0,$$

hence $|w|<1$; in other words, the interiors of the circles correspond.

The identical transformation $w = z$ is a special case of the above : if the point $z = 0$ corresponds to $w = 0$, then $a = 0$ and the transformation reduces to

$$w = ze^{i\theta}.$$

If, in addition, $dw/dz = 1$ when $z = 0$ we get

$$w = z.$$

III. The reader should find it easy to verify that *the transformation*

$$w = \frac{\rho(z-a)}{\rho^2 - \bar{a}z}$$

maps the circle $|z| = \rho$ *on the unit circle* $|w| = 1$. If $|a|<\rho$ it maps $|z|<\rho$ on $|w|<1$ and $|z|>\rho$ on $|w|>1$. If $|a|>\rho$ it maps $|z|>\rho$ on $|w|<1$ and $|z|<\rho$ on $|w|>1$.

IV. *Representation of the space bounded by three circular arcs on a rectilinear triangle.*

Consider three circles in the z-plane intersecting at the point $z = a$.

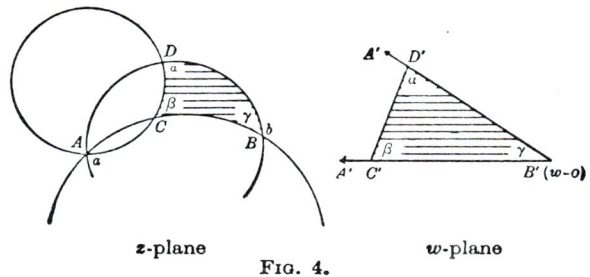

z-plane   w-plane

FIG. 4.

The angles of the curvilinear triangle BCD of fig. 4 are such that $\alpha + \beta + \gamma = \pi$. Consider the transformation

$$w = k\frac{z-b}{z-a} . . \qquad . \qquad . \qquad . \quad (1)$$

The point $A(z = a)$ corresponds to $w = \infty$ and $B(z = b)$ corresponds to $w = 0$. Equation (1) may be written

$$w - k = \frac{k(a-b)}{z-a},$$

or $w' = \lambda/z'$, where $w' = w - k$, $z' = z - a$ and $\lambda = k(a-b)$. Since the changes of variable from w to w' and from z to z' are mere translations, (1) is a pure inversion and reflexion. Since $z = a$ corresponds to $w = \infty$ and each of the three circular arcs BC, CD, DB passes through A they correspond to three straight lines in the w-plane. The two arcs BC, BD which pass through B correspond to two straight lines passing through $w = 0$, and the arc CD to a straight line which does not pass through $w = 0$.

It readily follows from (1) that the shaded areas of fig. 4 correspond.

With the usual convention of sign, we regard a motion round a closed simple contour, such as a circle, in the clockwise sense as positive for the area outside and negative for the area inside the contour.

If, by any conformal transformation, three points A, B, C on a closed contour in the z-plane correspond to the three points A', B', C' in the w-plane lying on the corresponding closed contour, then the interiors correspond if the points A', B', C' occur in the same (counter-clockwise) order as the points A, B, C.

We can see in this way that the shaded areas in fig. 4 correspond. It also follows that the curvilinear triangle formed by the arcs AC, CD, DA in the z-plane corresponds to the portion of the w-plane $A'C'D'A'$ where A' is the point at infinity.

EXAMPLES II

1. (i) Prove that, if $u = x^2 - y^2$, $v = -y/(x^2+y^2)$, both u and v satisfy Laplace's equation, but that $u+iv$ is not a regular function of z.

(ii) Show that the families of curves $u = $ const. $v = $ const. cut orthogonally if $u = x^2/y$, $v = x^2 + 2y^2$ but that the transformation represented by $u + iv$ is not conformal.

2. Prove that, if $w = x + iby/a$, $0 < a < b$, the inside of the circle $x^2 + y^2 = a^2$ corresponds to the inside of an ellipse in the w-plane, but that the transformation is not conformal.

3. Prove that, for the transformation $w^2 = (z - a)(z - \beta)$, the critical points are $z = a$, $z = \beta$, $z = \frac{1}{2}(a + \beta)$, $w = 0$, $w = \pm\frac{1}{2}i(a - \beta)$.

Show also that the condition that $z = \infty$ is *not* a critical point of the transformation $w = f(z)$ is that $\lim\limits_{z \to \infty} z^2 f'(z)$ must be finite and not zero.

4. If, by the inversion transformation $x = k^2\xi/\rho^2$, $y = k^2\eta/\rho^2$, $z = k^2\zeta/\rho^2$, where $r\rho = k^2$, $r^2 = x^2 + y^2 + z^2$, $\rho^2 = \xi^2 + \eta^2 + \zeta^2$, the twice-differentiable function $V(x, y, z)$ becomes $V_1(\xi, \eta, \zeta)$, prove that if

$$(\partial^2/\partial\xi^2 + \partial^2/\partial\eta^2 + \partial^2/\partial\zeta^2)\, V_1 = 0, \text{ then}$$
$$(\partial^2/\partial x^2 + \partial^2/\partial y^2 + \partial^2/\partial z^2)\, (V/r) = 0.$$

5. If $w = \cosh z$, prove that the area of the region of the w-plane which corresponds to the rectangle bounded by the lines $x = 0$, $x = 2$, $y = 0$, $y = \frac{1}{4}\pi$ is $(\pi \sinh 4 - 8)/16$.

6. If a is real and $0 < c < \pi$, find the area of the domain in the w-plane which corresponds by the transformation $w = e^z$ to the rectangle $a - c \leqslant x \leqslant a + c$, $-c \leqslant y \leqslant c$. Find the ratio of the areas of the two corresponding domains and prove that the ratio $\to e^{2a}$ as $c \to 0$.

7. Show that, if the function $w = f(z)$, regular in $|z| < R$, maps the circle $|z| = r < R$ on a rectifiable curve C in the w-plane, then the length of C is given by

$$\int_0^{2\pi} |f'(re^{i\theta})|\, r d\theta.$$

Show that the length of the curve into which the semicircular arc $|z| = 1$, $-\frac{1}{2}\pi \leqslant \arg z \leqslant \frac{1}{2}\pi$ is transformed by $w = 4/(1 + z)^2$ is $2\sqrt{2} + 2\log(1 + \sqrt{2})$. (See § **22**, equation (4)).

8. Find the Möbius' transformations which make the sets of points in the z-plane (i) a, b, c, (ii) 2, $1 + i$, 0 correspond to the points 0, 1, ∞ of the w-plane. In case (ii) show by sketches the domains of the w-plane and z-plane which correspond.

9. Find a Möbius' transformation which maps the circle $|z| \leqslant 1$ on $|w-1| \leqslant 1$ and makes the points $z = 0$, 1 correspond to $w = \frac{1}{2}$, 0 respectively. Is the transformation uniquely determined by the data ?

10. Find the transformation which maps the outside of the circle $|z| = 1$ on the half-plane $\mathbf{R}w \geqslant 0$, so that the points $z = 1$, $-i$, -1 correspond to $w = i$, 0, $-i$ respectively. What corresponds in the w-plane to (i) the lines $\arg z = $ const., $|z| \geqslant 1$, (ii) the concentric circles $|z| = r$, $(r > 1)$?

11. Prove that $w = (1+iz)/(i+z)$ maps the part of the real axis between $z = 1$ and $z = -1$ on a semicircle in the w-plane.

Find all the figures that can be obtained from the originally selected part of the axis of x by successive applications of this transformation.

12. Find what regions of the w-plane correspond by the transformation $w = (z-i)/(z+i)$ to (i) the interior of a circle of centre $z = -i$, (ii) the region $y > 0$, $x > 0$, $|z+i| < 2$. Illustrate by diagrams. Show that the magnification is constant along any circle with $z = -i$ as centre.

13. Let C_1, $|z-z_1| = r_1$ and C_2, $|z-z_2| = r_2$ be two non-concentric circles in the z-plane, C_1 lying entirely within C_2. Show that, if $z = a$, $z = b$ are the limiting points of the system of coaxal circles determined by C_1 and C_2, then $w = k(z-b)/(z-a)$ transforms C_1 and C_2 into concentric circles in the w-plane with centres at $w = 0$. If the radii of these concentric circles are ρ_1 and ρ_2, show that, although there is an infinite number of such representations, $\rho_1 : \rho_2$ is a constant.

14. Prove that, if $w = (az+\beta)/(\gamma z+\delta)$ and $a\delta - \beta\gamma = 1$, then the linear and superficial magnifications are $|\gamma z+\delta|^{-2}$, $|\gamma z+\delta|^{-4}$.

Show that the circle $|\gamma z+\delta| = 1$ $(\gamma \neq 0)$ is the complete locus of points in the neighbourhood of which lengths and areas are unaltered by the transformation. Prove that lengths and areas within this circle are increased and lengths and areas outside this circle are decreased in magnitude by the transformation. (This circle is called the *isometric circle*.)

CHAPTER III

SOME SPECIAL TRANSFORMATIONS

§ 19. Introduction

The fundamental problem in the theory of conformal mapping is concerned with the possibility of transforming conformally a given domain D of the z-plane into any given domain D' of the w-plane. It is sufficient to consider whether it is possible to map conformally any given domain on the interior of a *circle*. For if $\zeta = f(z)$ maps D on $|\zeta| < 1$ and $w = F(\zeta)$ maps D' on $|\zeta| < 1$, then $w = F\{f(z)\}$ provides a conformal transformation of D into D'.

The fundamental existence theorem of Riemann states that *any region with a suitable boundary can be conformally represented on a circle by a biuniform transformation.* Rigorous proofs of this existence theorem are long and difficult, and it is beyond our scope to discuss the question here.

In the applications of conformal transformation to practical problems, the problem to be solved is as follows: given two domains D and D' with specified boundaries, find the function $w = f(z)$ which will transform D into D' so that the given boundaries correspond. Although, by Riemann's existence theorem, we can infer the existence of the regular function $f(z)$, the theorem does not assist us to find the particular function $f(z)$ for each problem whose solution is desired. We have seen that when the two domains D and D' are bounded by *circles*, it is fairly easy to find the Möbius' transformation which maps D biuniformly on D'. Since for any arbitrary boundary curves there is no general method of finding the appropriate regular function $f(z)$, it is important to know the types of domain which correspond to each other when $f(z)$ is one of the *elementary* functions or a combination of several such functions.

In this chapter we discuss some of the most useful transformations which can be effected by elementary functions. The reader, who is mainly interested in the application of these transformations to practical problems, will find the special transformations discussed here of great value, but he must refer to other treatises for the details of the practical problems to which they can be applied.

Many useful transformations are obtained by combining several simple transformations.

For example, the transformation

$$w = \frac{(1+z^3)^2 - i(1-z^3)^2}{(1+z^3)^2 + i(1-z^3)^2}, \qquad \bullet \qquad \bullet \qquad \bullet \quad (1)$$

seems at first sight somewhat complicated, but on examination it is seen to be a combination of the successive simple transformations, $Z = z^3$, $\zeta = \dfrac{1+Z}{1-Z}$, $t = \zeta^2$, $w = \dfrac{t-i}{t+i}$.

It can be shown that (1) maps the circular sector $|z| < 1$, $0 < \arg z < \tfrac{1}{3}\pi$, conformally on the unit circle * $|w| < 1$.

§ 20. The Transformations $\mathbf{w = z^n}$

Let $w = u + iv = \rho e^{i\phi}$, $z = x + iy = r e^{i\theta}$, then it follows at once that $\rho = r^n$, $\phi = n\theta$, so that

$$u + iv = r^n (\cos n\theta + i \sin n\theta).$$

From the equations

$$u = r^n \cos n\theta, \; v = r^n \sin n\theta, \qquad \bullet \qquad \bullet \quad (1)$$

---

* **Compare § 24** IV, equation (8), with $a = \tfrac{1}{3}$.

either θ or r may be eliminated, giving

$$u^2 + v^2 = r^{2n} = (x^2 + y^2)^n, \qquad \bullet \qquad \bullet \qquad (2)$$

or
$$\tan n\theta = \frac{v}{u}. \qquad \bullet \qquad \bullet \qquad \bullet \qquad (3)$$

Equation (2) shows that the circles $r = c$ of the z-plane and the circles $\rho = c^n$ correspond, and in particular, that points on the circle $r = |z| = 1$ are transformed into points in the w-plane at unit distance from the origin. The lines $\theta = $ const., radiating from the origin of the z-plane, are transformed into similar radial lines $\phi = $ const. It should be noticed, however, that the line whose slope is θ in the z-plane is transformed into the line whose slope is $n\theta$ in the w-plane. Since $z = 0$ is a critical point of the transformation, the conformal property does not hold at this point.

In the simple case $w = z^2$, the angle between two radial lines in the z-plane is doubled in the w-plane. The case $w = z^2$ is typical, and we shall now consider it in greater detail.

We consider first the important difference between the transformation $w = z^2$ and the Möbius' transformations discussed in the preceding chapter. In the latter, points of the z-plane and of the w-plane were in one-one correspondence. For $w = z^2$, to each point z_0 there corresponds one and only one point $w_0 = z_0^2$, but to a point w_0 there correspond two values of z, $z = |\sqrt{w_0}|$, $z = -|\sqrt{w_0}|$. If we wish to preserve the one-one correspondence between the two planes, we may either consider the w-plane as slit along the real axis from the origin to infinity, or else construct the Riemann surface in the w-plane corresponding to the two-valued function of w defined by $w = z^2$. The method of constructing the Riemann surface was described in § **9**.

If we use the cut w-plane, then the upper half of the z-plane corresponds to the whole cut w-plane. There is a one-one correspondence between points of the upper half of the z-plane and points of the whole w-plane, and a one-one correspondence between points of the lower half of the z-plane with points of the whole w-plane ; but if we choose one of the two branches of $w = z^2$, say w_1, the cut plane effectively prevents our changing over, without knowing it, to the other branch w_2. The positive real z-axis corresponds to the upper edge and the negative real z-axis to the lower edge of the cut along the positive real axis in the w-plane.

If we use the two-sheeted Riemann surface in the w-plane, the sheet P_1 corresponds to the upper half of the z-plane for the branch w_1, and the sheet P_2 corresponds to the lower half of the z-plane for the branch w_2. Thus there is a one-one correspondence between the whole z-plane and the two-sheeted Riemann surface in the w-plane.

For $w = z^n$, where n is a positive integer, a wedge of the z-plane of angle $2\pi/n$ corresponds to the whole of the w-plane. If we divide up the z-plane into n such wedges, each of these corresponds to one of the n sheets of the n-sheeted Riemann surface in the w-plane.

If, for $w = z^2$, we cut the w-plane along the *negative* real axis, then the sheet P_1 of the Riemann surface corresponds to the half-plane $\mathbf{R}z \geqslant 0$, and the sheet P_2 to the half-plane $\mathbf{R}z \leqslant 0$.

§ 21. Further Consideration of $w = z^2$

From the equations

$$w = u + iv = (x + iy)^2 = x^2 - y^2 + 2ixy,$$

we have

$$u = x^2 - y^2, \quad v = 2xy. \tag{1}$$

By regarding u and v as curvilinear coordinates of points in the z-plane, the transformation $w = z^2$ can be examined from a knowledge of the curves in the z-plane which correspond to constant values of u and v. This method is frequently used in applying the theory of conformal transformation to practical problems.

Equations (1) show that the curves $u =$ const., $v =$ const. in the z-plane are two orthogonal families of rectangular hyperbolas.

The reader will easily verify that the shaded area in the z-plane of fig. 5 between two hyperbolas $2xy = v_1$,

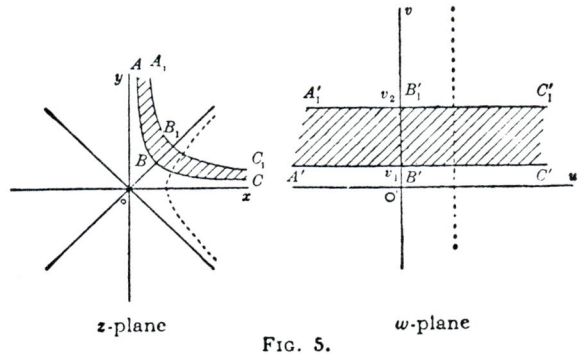

z-plane   w-plane

Fig. 5.

$2xy = v_2$ corresponds to the infinite strip of the w-plane shaded in the figure. Hence $w = z^2$ *maps the region between two hyperbolas on a parallel strip.*

If B_1 is at infinity, the point B'_1 is also at infinity, and the interior of the hyperbola ABC is transformed into the part of the upper half-plane above the line A' B' C'.

We also observe that *the transformation $w = z^2$ makes circles* $|z-a| = c$, $(a, c$ *real), in the z-plane correspond to limaçons in the w-plane.*

Consider the circle

$$z-a = ce^{i\theta}, \qquad . \qquad . \qquad . \qquad (2)$$

then

$$w-a^2+c^2 = 2c(c \cos \theta +a)e^{i\theta}.$$

Hence, on writing $w - a^2 + c^2 = Re^{i\theta}$, so that the pole in the w-plane is at $w = a^2 - c^2$, the polar equation of the curve into which (2) is transformed is the limaçon

$$R = 2ac + 2c^2 \cos \theta.$$

When $a = c$ the limaçon becomes a cardioid. This is the case if the circle (2) touches Oy at the origin.

§ 22. The Transformation $\mathbf{w} = \sqrt{\mathbf{z}}$

From the equations

$$u^2 - v^2 = x, \ 2uv = y, \qquad \bullet \qquad \bullet \qquad (1)$$

we get

$$y^2 = 4u^2(u^2 - x), \ y^2 = 4v^2(v^2 + x). \ . \qquad \bullet \qquad (2)$$

By means of the first of the equations (2), to the straight lines $u = $ const. correspond parabolas with vertex at $x = u^2$ and focus at the origin of the z-plane. To the orthogonal system of straight lines $v = $ const., we see, by the second of the equations (2), there corresponds another system of confocal parabolas with vertex at $x = -v^2$.

Consider the particular parabola of the first system corresponding to the value $u = 1$,

$$y^2 = 4(1 - x). \qquad \bullet \qquad \bullet \qquad \bullet \qquad (3)$$

Its transform in the w-plane is the line through the point $w = 1$ parallel to the v-axis. The points, A, B, C in fig. 6 correspond, in that order, to the points A', B', C'. The reader can easily verify that the shaded areas correspond, the two parabolas drawn corresponding to values $u = 1$, and $u = u_0 (>1)$.

If $u_0 \to \infty$, the region developed in the z-plane is the area outside the parabola (3), which accordingly corresponds to that part of the w-plane to the right of the line $u = 1$.

If the parameter u tends to zero, the parabola $y^2 = 4u^2(u^2 - x)$ narrows down until it becomes a slit along the negative real axis OX_1, which is a branch-line.

Hence the portion of the w-plane between the line $u = 1$ and the line $u = 0$ corresponds to the portion of the z-plane between the parabola ABC, $y^2 = 4(1-x)$, and the cut along the negative real axis OX_1 from the origin to $-\infty$.

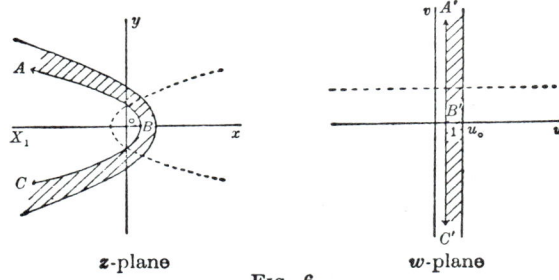

z-plane   w-plane

FIG. 6.

Hence we see that *the portion of the w-plane $\mathbf{R}w \geqslant 0$, corresponds to the whole z-plane cut along the negative real axis from 0 to $-\infty$.*

The simple w-plane is associated in a one-one correspondence with a two-sheeted Riemann surface covering the z-plane. The two sheets of the Riemann surface would be connected along the edges of the cuts along the negative real axis of the z-plane in the usual way.

The line $u = -1$ plainly corresponds to the same parabola $y^2 = 4(1-x)$ as does the line $u = 1$. Hence the portion of the w-plane to the left of the line $u = -1$ corresponds to the region outside the parabola ABC which lies on the second sheet of the Riemann surface.

If we combine $w = \sqrt{z}$ with a Möbius' transformation by writing $\zeta = (2/w) - 1$ we see that the transformation

$$\zeta = \frac{2}{\sqrt{z}} - 1 \qquad . \qquad . \qquad . \qquad (4)$$

transforms the region outside the parabola (3) *into the interior of the unit circle in the ζ-plane.* The points $z = 1$, $z = 4$,

$z = \infty$ correspond to the points $\zeta = 1$, $\zeta = 0$, $\zeta = -1$. The focus of the parabola (3), $z = 0$, lies *outside* the region of the z-plane which is under consideration, and it corresponds to the point $\zeta = \infty$ *outside* the unit circle $|\zeta| = 1$.

The reader should observe, however, that the preceding transformation cannot be used to represent the *inside* of the parabola *ABC* on the *inside* of the unit circle.

The transformation $w = \sqrt{z}$ just considered, illustrates an important point in the use of many-valued functions for solving problems in applied mathematics. The transformation $w = \sqrt{z}$ could be used to deal with a potential problem in which the field was the region *outside* the parabola *ABC* of fig. 6, but it could not be used for a problem in which the field was the space *inside* this parabola, since two points close to each other, one on each edge of the cut along the branch-line OX_1, will transform into two points on the axis of v, one in the upper and the other in the lower half-plane. Since these points are not close together in the w-plane, they would correspond to different potentials. It is important to realise that we cannot solve potential problems by using transformations which require a branch-line to be introduced into that part of the plane which represents the field.

§ 23. The Transformation $w = \tan^2(\frac{1}{4}\pi\sqrt{z})$

We have just seen that $w = (2/\sqrt{z}) - 1$ cannot be used to map the region *inside* the parabola $y^2 = 4(1-x)$ on the unit circle $|w| \leqslant 1$. We now consider a transformation which enables us to do this.

The transformation can be considered as a combination of the three transformations

$$w = \tan^2\tfrac{1}{2}\zeta, \quad \zeta = \tfrac{1}{2}\pi t, \quad t = \sqrt{z}.$$

where $w = u+iv$, $\zeta = \xi+i\eta$, $t = \sigma+i\tau$, $z = x+iy$.

The first transformation can be written

$$w = \frac{1-\cos\zeta}{1+\cos\zeta}.$$

If we consider the infinite strip between the lines $\xi = 0$, $\xi = \frac{1}{2}\pi$ of the ζ-plane, we see that, by writing $\zeta = \frac{1}{2}\pi + i\eta$, $\cos\zeta = -i\sinh\eta$ and $|w| = 1$. Thus, as y goes from $-\infty$ to ∞ along the line $\xi = \frac{1}{2}\pi$, w describes the unit-circle once. By writing $\zeta = i\eta$, $\cos\zeta = \cosh\eta$ and w is real. Thus as η goes from $+\infty$ to 0, w goes from -1 to 0 ; and as η goes from 0 to $-\infty$, w retraces its path from 0 to -1. Thus the strip $\xi = 0$, $\xi = \frac{1}{2}\pi$ corresponds to the cut-circle as illustrated in fig. 7. It is easy to verify

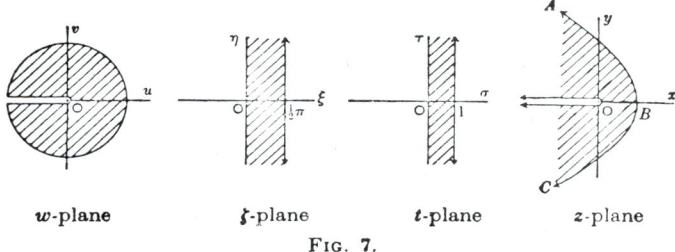

w-plane   **ζ-plane**   **t-plane**   **z-plane**

FIG. 7.

that the interiors correspond. The strip in the t-plane is plainly that between the lines $\sigma = 0$ and $\sigma = 1$. As we have already seen, $t = \sqrt{z}$ transforms the strip in the t-plane into the region inside the parabola ABC, $y^2 = 4(1-x)$, with a cut from the origin to infinity along the negative real axis. In fact, as $\sigma \to 0$ the parabola $y^2 = 4\sigma^2(\sigma^2 - x)$ becomes a very narrow parabola which is the slit illustrated in the z-plane in fig. 7.

The transformation $w = \tan^2(\frac{1}{4}\pi\sqrt{z})$ represents the region inside the parabola ABC on the inside of the unit circle $|w| = 1$ in a one-one correspondence, for the real axis of the w-plane between -1 and 0 corresponds to the real axis of the z-plane between $-\infty$ and 0. The cuts in the z-plane and w-plane are not needed for the *direct* transformation from the w-plane to the z-plane, but they are needed for the subsidiary transformations used in order to show how the boundaries of the various regions correspond.

Since $dw/dz = \pi \tan(\frac{1}{4}\pi\sqrt{z}) \sec^2(\frac{1}{4}\pi\sqrt{z})/4\sqrt{z}$, which tends to a finite non-zero limit as $z \to 0$, the points $z = 0$ and $w = 0$ are not critical points of the transformation, and so the representation is conformal as well as one-one.

§ 24. Combinations of $w = z^{\alpha}$ with Möbius' Transformations

I. *Semicircle on half-plane or circle.*

Consider the transformation

$$w = \left(\frac{z-ic}{z+ic}\right)^2; \quad (c \text{ real}). \qquad . \qquad . \qquad (1)$$

This is clearly a combination of

$$w = \zeta^2 \quad \text{and} \quad \zeta = (z-ic)/(z+ic).$$

The second of these may be written *

$$z = -ic\,\frac{\zeta+1}{\zeta-1}$$

for which it is clear that the circle $|z| = c$ corresponds to the imaginary axis of the ζ-plane $|\zeta+1| = |\zeta-1|$.

The boundary of the semicircle in the z-plane $ADCBA$ plainly corresponds to $A'D'C'B'A'$ in the ζ-plane, C' being the point $\zeta = -\infty$. The sense of description of the two boundaries shows that the shaded areas correspond.

Now consider $w = \zeta^2$: if $w = \rho e^{i\phi}$, $\zeta = re^{i\theta}$, we have

$$\rho = r^2, \quad \phi = 2\theta.$$

The shaded domain of the ζ-plane corresponds to $\pi \leqslant \theta \leqslant 3\pi/2$ and so the domain of the w-plane corresponding to this is $2\pi \leqslant \phi \leqslant 3\pi$, which is of course the same as $0 \leqslant \phi \leqslant \pi$, or the upper half of the w-plane.

---

* The use of the results of § **16** is frequently simpler than the procedure of splitting up the transformation into its real and imaginary parts.

Hence the interior of the shaded semicircle of fig. 8 corresponds to the upper half of the w-plane.

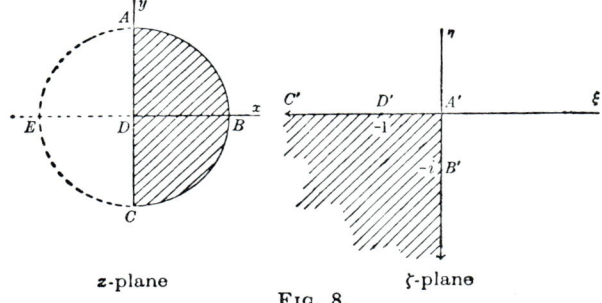

z-plane   ζ-plane

FIG. 8.

It is easy to verify that the upper half of the *w*-plane corresponds to the interior of the respective semicircles *BAEDB, AECDA, ECBDE* by the transformations

$$w = \left(\frac{z+c}{z-c}\right)^2, \ w = \left(\frac{z+ic}{z-ic}\right)^2, \ w = \left(\frac{z-c}{z+c}\right)^2.$$

Also, by combining (1) with the transformation

$$t = \frac{i-w}{i+w},$$

the transformation

$$t = \frac{i - \left(\dfrac{z-ic}{z+ic}\right)^2}{i + \left(\dfrac{z-ic}{z+ic}\right)^2} = i\,\frac{z^2-c^2+2cz}{z^2-c^2-2cz}$$

conformally represents the interior of the z-semicircle ABCDA on the interior of the unit circle $|\,t\,| = 1$.

II. *Wedge or sector on half-plane.*

By the transformation

$$w = z^{1/a}, \qquad \bullet \qquad \bullet \qquad \bullet \qquad (2)$$

the area bounded by the infinite wedge of angle πa with its vertex at $z = 0$ and one arm of the angle along the positive x-axis is transformed into the upper half of the w-plane.

The reader will find this quite easy to verify.

The sector cut off from this wedge by an arc of the unit-circle $|z| = 1$ is transformed by (2) into the unit semicircle in the upper half of the w-plane.

This is also easy to verify.

III. *Circular crescent or semicircle on half-plane.*

We readily see, from § **18**, IV, that the circular crescent with its points at $z = a$ and $z = b$ and whose angle is πa can be transformed into the wedge mentioned in II above by

$$\zeta = k\,\frac{z-a}{z-b}$$

if the constant **k** be suitably chosen. *Hence the crescent can be transformed into the w-half-plane by*

$$w = c\left(\frac{z-a}{z-b}\right)^{1/a} \qquad . \qquad . \qquad . \quad (3)$$

A semicircle may be regarded as a particular case of a crescent in which $a = \frac{1}{2}$. The semicircle of radius unity and centre $z = 0$ lying in the upper half-plane is transformed into the first quadrant of the ζ-plane by

$$\zeta = \frac{1+z}{1-z} \qquad . \qquad . \qquad . \qquad . \quad (4)$$

The quarter-plane becomes a half-plane by * $w = \zeta^2$ and so *the semicircle is transformed into the upper half of the w-plane by*

$$w = \zeta^2 = \left(\frac{1+z}{1-z}\right)^2 \qquad . \qquad . \qquad , \quad (5)$$

---

* See I above.

IV. *Sector on unit circle.*

Consider the sector in the z-plane, shaded in fig. 9. Let us find the transformation which represents this sector on a unit circle.

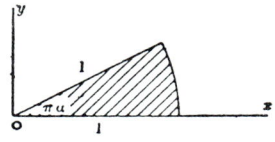

FIG. 9.

By means of (2) the sector is transformed into the unit-semicircle in the upper half of the w-plane. By means of (5) we see that this unit semicircle is transformed into the upper half of the t-plane by

$$t = \left(\frac{1+w}{1-w}\right)^2. \qquad \bullet \qquad \bullet \qquad \bullet \qquad (6)$$

Again, the upper half of the t-plane is transformed into the interior of the unit circle in the ζ-plane, $|\zeta| = 1$, by

$$\zeta = \frac{t-i}{t+i} \quad . \qquad \bullet \qquad \bullet \qquad \bullet \qquad (7)$$

and on combining these, *the transformation which represents the shaded area of fig. 9 on the unit circle in the ζ-plane is* *

$$\zeta = \frac{\left(\dfrac{1+z^{1/a}}{1-z^{1/a}}\right)^2 - i}{\left(\dfrac{1+z^{1/a}}{1-z^{1/a}}\right)^2 + i} \quad \bullet \qquad \bullet \qquad \bullet \qquad (8)$$

V. By combining $w = \sqrt{z}$ with a Möbius' transformation we find in a similar way *the transformation which represents the z-plane, cut from 0 to ∞ along the positive real axis on the unit circle $|\zeta| \leqslant 1$* in the form

$$\zeta = \frac{\sqrt{z}-i}{\sqrt{z}+i}.$$

* When $a = \frac{1}{2}$, this transformation is the same as (1) of § **19**.

VI. *Transformations of the cut-plane.*

Consider the two transformations

$$w = \frac{z-a}{z-b}, \; w = -\frac{z-a}{z-b}$$

where a and b are real and $a > b$. By means of the first of these, the z-plane, cut along the real axis from $z = a$ to $+\infty$ and from $z = b$ to $-\infty$, is transformed into the w-plane cut from $w = 0$ to $w = \infty$, the cut passing through the point $w = 1$ which corresponds to $z = \infty$. By means of the second, the z-plane cut from $z = a$ to $z = b$ is transformed into the w-plane cut along the *positive* real axis from 0 to ∞. The cut in this case does *not* pass through $w = -1$, the point corresponding to $z = \infty$.

§ 25. Exponential and Logarithmic Transformations

Most of the transformations so far considered have been Möbius' transformations, $w = z^a$ and combinations of these two types. We now observe that the relation

$$w = e^z \qquad . \qquad . \qquad . \qquad (1)$$

gives rise to two important special transformations.

If we use rectangular coordinates x, y and polar coordinates ρ, ϕ in the w-plane we get

$$\rho = e^x, \; \phi = y.$$

The *horizontal strip* of the z-plane bounded by the lines $y = y_1$ and $y = y_2$ where $|y_1 - y_2| < 2\pi$ is transformed into a wedge-shaped region of the w-plane, the angle of the wedge being $\alpha = |\phi_2 - \phi_1| = |y_2 - y_1|$. The representation is conformal throughout the interior of these regions since dw/dz is never zero. In particular, if $y_1 = 0$, $y_2 = \pi$, so that $|y_2 - y_1| = \pi$, the wedge becomes a half-plane. The *semi*-infinite strip $-\infty \leqslant x \leqslant 0$, $0 \leqslant y \leqslant \pi$ is readily seen to correspond to unit semicircle in the upper half of the w-plane.

If $|y_1 - y_2| > 2\pi$, the wedge obtained covers part of the

w-plane multiply. We may in this case make use of the cut w-plane. If $y_1 = 0$, $y_2 = 2\pi$, the strip of width 2π in the z-plane corresponds to the w-plane cut along the positive real axis. When $|y_1 - y_2|$ is an integral multiple of 2π the strip is transformed into a Riemann surface. Each strip of the z-plane of breadth 2π corresponds to one sheet of the ∞-sheeted Riemann surface.

A second special transformation is obtained from (1) by considering an arbitrary *vertical strip* bounded by the lines $x = x_1$, $x = x_2$, $(x_1 < x_2)$. This strip is represented on a Riemann surface which covers the annulus between the concentric circles $|w| = \rho_1$, $|w| = \rho_2$ an infinite number of times. If we keep x_2 constant and let $x_1 \to -\infty$, the strip $x_1 < x < x_2$ becomes in the limit the portion of the z-plane to the left of the line $x = x_2$, and we obtain in the w-plane a Riemann surface which covers the circle $|w| < \rho_2$ except at the point $w = 0$, where it has a logarithmic branch-point.

The inverse function

$$w = \operatorname{Log} z . \qquad . \qquad . \qquad . \qquad (2)$$

gives, on interchanging the z-plane and w-plane, exactly the same transformations as (1).

It should be remembered that although $\operatorname{Log} z$ is an infinitely many-valued function of z, e^z is one-valued.

Since $z^a = e^{a \operatorname{Log} z}$ the transformation $w = z^a$ may be regarded as a combination of the two transformations

$$w = e^\zeta , \ \zeta = a \operatorname{Log} z.$$

§ 26. Transformations involving Confocal Conics

Consider the transformation

$$2z = (a-b)w + \frac{a+b}{w} . . \qquad . \qquad . \qquad (1)$$

If $w = re^{i\theta}$, we get

$$2x = \left\{ (a-b)r + \frac{a+b}{r} \right\} \cos \theta, \ 2y = \left\{ (a-b)r - \frac{a+b}{r} \right\} \sin \theta,$$

and so the curves in the z-plane, corresponding to concentric circles in the w-plane having the origin for their centre, are confocal ellipses, the distance between the foci being $2\sqrt{(a^2-b^2)}$. The curves in the z-plane corresponding to straight lines through the origin in the w-plane are the confocal hyperbolas, a result to be expected, since the two families of curves in each plane must cut orthogonally. Clearly there is no loss of generality by taking $a = 1$, $b = 0$, and so we may consider the transformation

$$2z = w + \frac{1}{w} \qquad . \qquad . \qquad . \qquad (2)$$

as typical. Clearly z becomes infinite when $w = 0$, and since

$$\frac{dz}{dw} = \frac{1}{2}\left(1 - \frac{1}{w^2}\right),$$

the points $w = \pm 1$, at which the derivative vanishes, are critical points of the transformation. We now have

$$2x = \left(r + \frac{1}{r}\right)\cos\theta, \; 2y = \left(r - \frac{1}{r}\right)\sin\theta$$

and on eliminating θ, we get the ellipse in the z-plane

$$\frac{x^2}{\frac{1}{4}\left(r + \frac{1}{r}\right)^2} + \frac{y^2}{\frac{1}{4}\left(r - \frac{1}{r}\right)^2} = 1, \qquad . \qquad . \qquad (3)$$

corresponding to each of the two circles $|w| = r$, $|w| = 1/r$. As $r \to 1$, the major semi-axis of the ellipse tends to 1, while the minor semi-axis tends to zero. As $r \to 0$, or as $r \to \infty$, both semi-axes tend to infinity. From this it is plain that the inside and the outside of the unit circle in the w-plane *both* correspond to the whole z-plane, cut along the real axis from -1 to 1. The unit circle $|w| = 1$ itself corresponds to a very narrow ellipse, which is the cut along the real axis enclosing the critical points -1 and 1.

On solving equation (2) for w we get

$$w = z \pm \sqrt{(z^2 - 1)}$$

and the inverse function is a two-valued function of z. If we choose the lower sign, *the transformation*

$$w = z - \sqrt{(z^2 - 1)} \qquad . \qquad \bullet \qquad \bullet \qquad (4)$$

transforms the area outside the ellipse (3) *conformally into the* inside *of the circle* $|w| = r$. The lower sign is the correct one to select, since the point $w = 0$ inside the circle must correspond to the point $z = \infty$; the other sign of the square root would make the points $w = \infty$, $z = \infty$ correspond. The region between two confocal ellipses in the z-plane is transformed into the annulus between two concentric circles in the w-plane.

The function (4) *gives a transformation of the z-plane, cut along the real axis from* -1 *to* 1, *on the interior of the circle* $|w| = 1$.

If we take the other sign, it is clear that *the transformation*

$$w = z + \sqrt{(z^2 - 1)}$$

gives a transformation of the cut z-plane on the outside *of the unit circle* $|w| = 1$. The relation (2) is remarkable in that it represents the cut z-plane not only on the *interior* but also on the *exterior* of the unit circle $|w| = 1$.

The ambiguity can be removed from this transformation by replacing the z-plane by a Riemann surface of two sheets, each cut from -1 to 1 and joined crossways along the cut. Then, of course, the interior of the unit circle $|w| = 1$ corresponds to one sheet and the exterior of the unit circle to the other sheet of this Riemann surface.

If $r > 1$, the transformation (2) maps the exterior of the circle $|w| = r$, or the interior of the circle $|w| = 1/r$, on the *exterior* of the ellipse (3). It should be observed, however, that the *interior* of the ellipse cannot be represented on the *interior* of the unit circle by any of the elementary transformations so far employed. We may remark, however, that the upper half of the ellipse (3) is represented by (2) on the upper half of an annular region cut along the real axis : this last area, and hence the semi-ellipse also, can be transformed into a rectangle by a method similar to that described in § 25. The transformation which maps the *interior* of an ellipse on a unit circle involves elliptic functions.

§ 27. The Transformation z = c sin w

From the relation

$$z = c \sin w, \quad (c \text{ real}) \quad . \quad . \quad . \quad (1)$$

we get, on equating real and imaginary parts,

$$x = c \sin u \cosh v, \, y = c \cos u \sinh v,$$

so that, when v is constant, the point z describes the curves

$$\frac{x^2}{c^2 \cosh^2 v} + \frac{y^2}{c^2 \sinh^2 v} = 1, \quad . \quad . \quad (2)$$

which, for different values of v, are confocal ellipses. Consider a rectangle in the w-plane bounded by the lines $u = \pm \frac{1}{2}\pi, v = \pm \lambda$. For all values of u, cos u is positive ; hence when $v = \lambda$, y is positive and x varies from $c \cosh \lambda$ to $-c \cosh \lambda$, that is, the half of the ellipse on the positive side of the axis of x is covered.

Let $u = -\frac{1}{2}\pi$, then $y = 0$ and $x = -c \cosh v$. Hence as v varies from λ through zero to $-\lambda$ along the side of the rectangle, x passes from A' to the focus H (see fig. 10) and back from H to A'.

When $v = -\lambda$ then z describes the half of the ellipse on the negative side of the axis of x. When $u = \frac{1}{2}\pi$ then $y = 0$ and $x = c \cosh v$, so that z moves from A to the focus S and back from S to A.

Hence the curve in the z-plane corresponding to the contour of the rectangle in the w-plane is the ellipse with two slits from the extremities of the major axis each to the nearer focus. It is easy to see that the two interiors correspond.

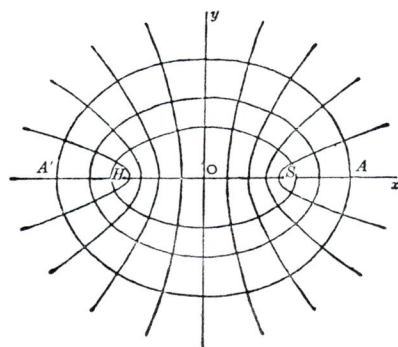

Fig. 10.

Since $\sin w = \cos(\tfrac{1}{2}\pi - w)$, the transformation given by

$$z = c \cos w$$

can be dealt with in a similar way. The details are left to the reader.

The function inverse to (1),

$$w = \arcsin(z/c),$$

is an infinitely many-valued function of z. If we use the cut plane, the cuts must be from S to infinity along the positive real axis and from H to infinity along the negative real axis. The Riemann surface of an infinite number of sheets in the z-plane, which would secure unique correspondence between every z-point and every w-point, would have the junctions of its different sheets along the above-mentioned cuts.

§ 28. Joukowski's Aerofoil

The transformation

$$\frac{w-\kappa c}{w+\kappa c} = \left(\frac{z-c}{z+c}\right)^{\kappa} \qquad . \qquad . \qquad . \qquad (1)$$

is important in the practical problem of mapping an aeroplane-wing profile on a nearly circular curve. If the profile has a sharp point at the trailing edge and we write $\beta = (2-\kappa)\pi$, then β is the angle between the tangents to the upper and lower parts of the profile at this point. If a circle is drawn through the point $-c$ in the z-plane, so that it just encloses the point $z = c$ and cuts the line joining $z = -c$ and $z = c$ at $z = c+\epsilon$ where ϵ is small, this circle is mapped by (1) on a wing-shaped curve in the w-plane.

A special case of (1) when $c = 1$, $\kappa = 2$ will now be discussed in detail.

In practical problems on the study of the flow of air round an aerofoil, the transformation desired is one which maps the region *outside* the aerofoil on the region *outside* a circle or nearly circular curve. The special case of (1) when $c = 1$, $\kappa = 2$,

$$\frac{w-2}{w+2} = \left(\frac{z-1}{z+1}\right)^{2} \qquad . \qquad . \qquad . \qquad (2)$$

transforms a circle in the z-plane, passing through the point $z = -1$ and containing the point $z = 1$, into a wing-shaped curve in the w-plane, known as *Joukowski's profile*.

We readily see that (2) is the same as the transformation, already discussed,

$$w = z + \frac{1}{z}. \qquad . \qquad . \qquad . \qquad (3)$$

If C is a circle in the z-plane passing through the point $z = -1$, such that the point $z = 1$ is within C, the transformation (3) maps the outside of C conformally on the outside of Joukowski's profile F. The shape of the curve F can easily be obtained from the circle C by making the point z trace out this circle, and adding the vectors z and $1/z$. See fig. 11.

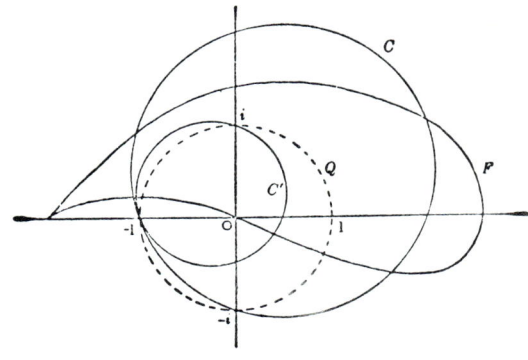

Fɪɢ. 11.

We may also consider (2) as a combination of the three transformations

$$t = \frac{z+1}{z-1}, \quad \zeta = t^2, \quad w = 2\,\frac{\zeta+1}{\zeta-1}.$$

By the first of these, the circle C is transformed into a circle Γ in the t-plane passing through $t = 0$. By the second, the circle Γ is transformed into a cardioid * in the ζ-plane with cusp at $\zeta = 0$. The third transformation then maps the cardioid on the wing-shaped curve F in the w-plane. Since $z = 1$ corresponds to $t = \infty$, the outside of C is mapped on the interior of Γ. The interior of the cardioid corresponds to the interior of Γ. Since $\zeta = 1$ corresponds to $w = \infty$, the outside of F corresponds to the inside of the cardioid, and so to the outside of C.

In fig. 11, C is the given circle, C' the circle obtained from C by the transformation $1/z$ and Q is the unit circle.

* See § **21.**

The critical points of (3) are $z = 1$ and $z = -1$, and since the point $z = 1$ is inside C, the mapping of the outside of C on the outside of F is conformal.

If the point $z = 1$ is outside C, then the inside of F corresponds to the inside of C, and the figure corresponding to this case is the same as fig. 11 with the circles C and C' interchanged.

§ 29. Some Important Transformations Tabulated

In Table 1 we tabulate for convenience a number of examples of domains in the z-plane which can be mapped conformally on the interior of the unit circle $|w| \leqslant 1$.

TABLE 1

	Domain in the z-plane $z = x + iy = re^{i\theta}$.	Domain in the w-plane $w = u + iv = \rho e^{i\phi}$.	Transformation.
1.	Unit circle $|z| \leqslant 1$	Unit circle $|w| \leqslant 1$	$w = e^{i\lambda} \dfrac{z-a}{\bar{a}z-1}$
2.	Upper half-plane	Unit circle $|w| \leqslant 1$	$w = \dfrac{i-z}{i+z}$
3.	Infinite strip of finite breadth $-\infty \leqslant y \leqslant \infty$, $0 \leqslant x \leqslant \pi$	Unit circle $|w| \leqslant 1$	$\dfrac{w-1}{w+1} = ie^{iz}$
4.	Area outside the ellipse $x^2/a^2 + y^2/b^2 = 1$	Unit circle $|w| \leqslant 1$	$2z = (a-b)w + \dfrac{a+b}{w}$
5.	Area outside the parabola $r \cos^2 \tfrac{1}{2}\theta = 1$	Unit circle $|w| \leqslant 1$	$(w+1)^2 z = 4$
6.	Area within same parabola	Unit circle $|w| \leqslant 1$	$w = \tan^2(\tfrac{1}{4}\pi\sqrt{z})$
7.	Semicircle $x^2 + y^2 = c^2,\ x > 0$	Unit circle $|w| \leqslant 1$	$w = i\dfrac{z^2 - c^2 + 2cz}{z^2 - c^2 - 2cz}$

Some useful conformal transformations in which the domain in the w-plane is not a circle are given in Table 2. When the domain in the w-plane is either the upper half-plane or a semicircle it can of course be transformed into a circle by either 2 or 7 of Table 1.

TABLE 2

	Domain in z-plane.	Domain in w-plane.	Transformation.
1.	The angle $0 \leqslant \theta < 2\pi/n$	Upper half-plane	$w = z^n$
2a.	Semicircle $\|z\| \leqslant 1, y > 0$	Upper half-plane	$w = \left(\dfrac{z+1}{z-1}\right)^2$
2b.	Semicircle $\|z\| \leqslant 1, y > 0$	First quadrant of w-plane	$w = \dfrac{1+z}{1-z}$
3.	Sector $\|z\| \leqslant 1, 0 \leqslant \theta < \pi/n$	Upper half-plane	$w = \left(\dfrac{z^n+1}{z^n-1}\right)^2$
4.	Domain between two circular arcs intersecting at $z = a$, $z = b$ at angles π/n	Half-plane	$w = \left(\dfrac{z-a}{z-b}\right)^n$
5.	Strip $0 \leqslant y \leqslant \pi$	Upper half-plane	$w = e^z$
6.	Outside of parabola $y^2 = 4c^2(x+c^2)$ (not containing the focus $z = 0$)	Upper part of plane above the line $v = c > 0$	$w = \sqrt{z}$
7a.	Upper half of the inside of the same parabola	Strip $0 \leqslant u \leqslant \infty, 0 < v < c$	$w = \sqrt{z}$
7b.	Strip $0 \leqslant x < \infty, 0 < y < c$	Semicircle $\|w\| < 1, v > 0$	$w = -e^{-\pi z/c}$
8.	Domain between the two branches of the hyperbola $\dfrac{x^2}{(2\cos\alpha)^2} - \dfrac{y^2}{(2\sin\alpha)^2} = 1$, $0 < \alpha < \tfrac{1}{2}\pi$	The angle (a) $\alpha < \phi < \pi - \alpha$, and (b) $\alpha + \pi < \phi < 2\pi - \alpha$	(a) $w = \dfrac{z + \sqrt{(z^2-4)}}{2}$ (b) $w = \dfrac{z - \sqrt{(z^2-4)}}{2}$
9.	Upper half of inside of the branch of the same hyperbola $x > 0, y > 0$	Sector $\|w\| < 1, -\alpha < \phi < 0$	$w = \dfrac{z - \sqrt{(z^2-4)}}{2}$
10.	Upper half of inside of ellipse $x^2/a^2 + y^2/b^2 = 1, y > 0$ $a = r+1/r, b = 1/r - r, r < 1.$	Annulus between $\|w\| = 1$, $\|w\| = r < 1, v > 0$	$w = \dfrac{z + \sqrt{(z^2-4)}}{2}$

It is impossible, in the limited space at our disposal, to discuss all the transformations which are of practical importance. It is important, however, to mention briefly the **Schwarz-Christoffel** transformation, which has numerous important applications.

Let $a, b, c, \ldots$ be n points on the real axis in the w-plane such that $a < b < c < \ldots$; and let $\alpha, \beta, \gamma, \ldots$ be interior angles of a simple closed polygon of n vertices so that $\alpha + \beta + \gamma + \ldots = (n-2)\pi$.

Then the transformation of Schwarz-Christoffel is a transformation from the w-plane to the z-plane defined by

$$\frac{dz}{dw} = K(w-a)^{\frac{\alpha}{\pi}-1}(w-b)^{\frac{\beta}{\pi}-1}(w-c)^{\frac{\gamma}{\pi}-1} \ldots .$$

It transforms the real axis in the w-plane into the boundary of a closed polygon in the z-plane in such a way that the vertices of the polygon correspond to the points $a, b, c, \ldots$ and the interior angles of the polygon are $\alpha, \beta, \gamma, \ldots$. When the polygon is simple, the interior is mapped by this transformation on the upper half of the w-plane. The number K is a constant, which may be complex.

If we write $K = Ae^{i\lambda}$, where A and λ are real, one vertex of the polygon can be made to correspond to the point at infinity on the w-axis. If $a \to -\infty$ we can choose A to be of the form $B(-a)^{-\frac{\alpha}{\pi}+1}$ and since, as $a \to -\infty$, $\{(w-a)/-a\}^{\frac{\alpha}{\pi}-1} \to 1$, the transformation becomes

$$\frac{dz}{dw} = Be^{i\lambda}(w-b)^{\frac{\beta}{\pi}-1}(w-c)^{\frac{\gamma}{\pi}-1} \ldots .$$

The reader who desires further information about this important transformation is referred to larger treatises.*

* See *e.g.* Copson, *Functions of a Complex Variable* (Oxford, 1935), p. 193 *seq.*

EXAMPLES III

1. Prove that, by the transformation

$$w = \frac{c}{2}\left(\frac{z-a}{z+a} + \frac{z+a}{z-a}\right),$$

two sets of coaxal circles are transformed into sets of confocal conics. What region of the w-plane corresponds to the inside of the circle $|(z-a)/(z+a)| = \frac{1}{2}$?

2. Show that by the transformation $w = \left(\frac{z+i}{z-i} + 1\right)^2$, the real axis in the z-plane corresponds to a cardioid in the w-plane. Indicate the region of the z-plane which corresponds to the interior of the cardioid.

3. If $w = -ic \cot \frac{1}{2}z$, where c is real, show that the rectangle bounded by $x = 0$, $x = \pi$, $y = 0$, $y = \infty$, is conformally represented on a quarter of the w-plane. Find a transformation $\zeta = f(z)$ which maps this infinite rectangle on the semicircle $|\zeta| \leqslant a$, $\xi \geqslant 0$.

4. If $w = \tan z$, prove that

$$u^2 + v^2 + 2u \cot 2x - 1 = 0, \quad u^2 + v^2 - 2v \coth 2y + 1 = 0.$$

Hence show that the strip $-\frac{1}{2}\pi < x < \frac{1}{2}\pi$ corresponds to the whole w-plane. To obtain a Riemann surface in the w-plane so as to secure unique correspondence between every w-point and every z-point, show that the w-plane must be cut along the imaginary axis from i to ∞ and from $-i$ to $-\infty$.

Investigate $w = \tan z$ as a combination of

$$iw = (\zeta-1)/(\zeta+1), \quad \zeta = e^{2iz}.$$

5. Find the curves in the z-plane corresponding to $|w| = 1$ if

$$w = \frac{z(z-\sqrt{2})}{1-\sqrt{2}z}.$$

6. Show that $w = 2z/(1-z^2)$ maps two of the four domains, into which the circles $|z-1| = \sqrt{2}$, $|z+1| = \sqrt{2}$ divide the z-plane, conformally on $|w| < 1$.

7. Prove that, if $3z^2 - 2wz + 1 = 0$, the annulus $1/\sqrt{3} < |z| < 1$ is mapped conformally on the interior of the ellipse $u^2 + 4v^2 = 4$ cut along the real axis between its foci. Discuss what corresponds in the w-plane to the curves (i) $|z| = r$, (ii) $\arg z = \alpha$.

8. Find the transformation which maps the outside of the ellipse $|z-2| + |z+2| = 100/7$ on the circle $|w| \leqslant 1$.

9. Show that $w = -\frac{1}{2}(z + 1/z)$ maps the upper half of the circle $|z| < 1$ on the upper half of the w-plane. At what points of the z-plane is the linear magnification equal to $\frac{1}{2}$? At what points is the rotation equal to $\pm\frac{1}{2}\pi$? Prove that the magnification is greater than unity throughout the interior of the semicircle $3|z|^2 = 1$ in the upper half-plane.

10. By considering the successive transformations $\zeta = -\frac{1}{2}(z + 1/z)$, $w = 1/\zeta^2$ prove that $w = 4z^2/(1+z^2)^2$ maps the upper half of the circle $|z| < 1$ on the w-plane, cut along the positive real axis, so that the points $z = 0, 1, i$ correspond to $w = 0, 1, \infty$ respectively.

What points of the w-plane correspond to $z = \pm 1$?

11. Show that by $w = e^{\pi z/a}$ an equiangular spiral in the w-plane corresponds to a straight line in the z-plane.

12. Discuss the transformation

$$w = \log\left\{ \frac{\sqrt{(z-\alpha)} + \sqrt{(z-\beta)}}{\sqrt{(\beta-\alpha)}} \right\},$$

showing that the lines $u = \text{const.}$, $v = \text{const.}$ correspond to sets of confocal conics with foci at $z = \alpha$, $z = \beta$.

13. Show that $2w = \log\{(1+z)/(1-z)\}$ represents $|z| < 1$ on the strip of the w-plane $-\frac{1}{4}\pi < v < \frac{1}{4}\pi$.

14. Show that $iw = \log\{\sqrt{(z/a)} - 1\}$ represents the strip $v = 0$, $v = \infty$, $u = -\pi$, $u = \pi$ on the interior of the cardioid $r = 2a(1 + \cos\theta)$ in the z-plane cut along the real axis from the cusp to $x = a$.

15. Show that, if c is real,

$$\frac{z-c}{z+c} = ie^{iw}$$

conformally transforms the strip $v = -\infty$, $v = \infty$, $u = 0$, $u = \pi$, into the circle $|z| \leqslant c$.

16. Show that, by the relation $w^2 = 1 + e^z$, the lines $x = \text{const.}$ are transformed into a series of confocal lemniscates (Cassini's ovals) in the w-plane.

If $a>1$ and $z^2(a^2+w^2-1) = aw^2$, show that the interior of the circle $|z| = 1$ is transformed into the interior of the Cassini's oval $\rho\rho' = a$, where ρ and ρ' are the distances of a point from the foci $(1,\ 0)$ and $(-1,\ 0)$.

17. If $z = x+iy$, prove that the inside of the parabola $y^2 = 4c^2(x+c^2)$ is mapped on the upper half of the w-plane by

$$w = i \cosh \frac{\pi\sqrt{z}}{2c}.$$

18. Show that the transformation

$$\frac{z}{c} = \left(\frac{1+w}{1-w}\right)^{2a/\pi i}$$

transforms the inside of the circle $|w| = 1$ with two semi-circular indentations, of centres 1 and -1, drawn so as to exclude these points from the circular area and boundary, into the annulus between two circles in the z-plane, of centre the origin and radii ce^a, ce^{-a}, with a single slit along the real axis.

19. If $0 \leqslant a < \beta < 2\pi$, show that

$$w = (ze^{-ia})^{2\pi/(\beta-a)}$$

maps the region $a < \arg z < \beta$ on the w-plane cut along the positive real axis. Hence find the transformation $w = f(z)$ which maps the circular sector $a < \arg z < \beta$, $|z| < 1$, on the circle $|w| < 1$.

20. Use the successive transformations

$$\zeta = (z+\tfrac{1}{2})\pi i, \ \ s = e^{\zeta}, \ \ t = \frac{1+s}{1-s}, \ \ r = t^2, \ \ w = \frac{r-i}{r+i},$$

to form the single transformation $w = f(z)$ which maps the strip $-\tfrac{1}{2} \leqslant x \leqslant \tfrac{1}{2}$, $y \geqslant 0$ of the z-plane on $|w| \leqslant 1$.

21. Use the transformations $\zeta = \sqrt{z}, t = \sin \tfrac{1}{2}\pi\zeta, w = \dfrac{t-1}{t+1}$, to show that

$$w = \frac{\sin \tfrac{1}{2}\pi\sqrt{z}-1}{\sin \tfrac{1}{2}\pi\sqrt{z}+1}$$

maps the inside of the parabola $r = 2/(1+\cos\theta)$ in the z-plane, cut from the focus ($z = 0$) to the point $z = -\infty$, on the unit circle in the w-plane cut from $w = 0$ to $w = 1$.

22. Find the equations of the curves in the z-plane which correspond to constant values of u and v if $z = w + e^w$. What corresponds to the lines $v = 0$, $v = \pi$? Sketch some of the curves $v = $ const. for values of v between $-\pi$ and π.

23. Show that the transformation

$$w/a = i \sinh \tfrac{1}{2}(z - i\beta)/\cosh \tfrac{1}{2}(z + i\beta)$$

transforms one of the regions bounded by the orthogonal circles $|w| = a$ and $|w - a \operatorname{cosec} \beta| = a \cot \beta$ into the infinite strip $0 \leqslant y \leqslant \tfrac{1}{2}\pi$.

24. If $w = \tanh z$, show that the lines $x = $ const., $y = $ const., correspond to coaxal circles in the w-plane.

Prove that this transformation maps the strip $0 < y < \tfrac{1}{4}\pi$ conformally on the upper-half of the w-plane.

25. Prove that, if $0 < c < 1$, the transformation

$$w = \frac{z(z - c)}{cz - 1}$$

transforms the unit circle in the z-plane into the unit circle, taken twice, in the w-plane, and the inside of the first circle into the inside, taken twice, of the second.

26. Prove that, if $a > 0$, $h > 0$,

$$w = 2a\sqrt{(z + 1)} + \log \frac{\sqrt{(z + 1)} + 1}{\sqrt{(z + 1)} - 1} + i\pi$$

maps the upper-half of the z-plane on the positive quadrant of the w-plane with a slit along the line $v = \pi$, $u \geqslant h$, where $w = h + i\pi$ when $z = 1/a$. (See p. 80.)

27. Show that by the transformation

$$\frac{dz}{dw} = \frac{1}{\sqrt{w}\sqrt{(1 - w^2)}}$$

the upper half of the w-plane can be mapped on the interior of a square, the length of a side of which is

$$\int_0^{\frac{1}{2}\pi} \sqrt{(\operatorname{cosec} \phi)}\, d\phi.$$

CHAPTER IV

THE COMPLEX INTEGRAL CALCULUS

§ 30. Complex Integration

The development of the theory of functions of a complex variable follows quite a different line from that of functions of a real variable. In the latter theory, having discussed functions which possess a derivative, we proceed to consider the more special class of functions which possess derivatives of the second order ; then, from among those functions which possess derivatives of all orders, we select those which can be expanded in a power series by Taylor's theorem. In complex variable theory, on the other hand, we begin by dealing with regular functions, and, by virtue of the definition of regularity, the class of functions is so restricted that a function which is regular in a region possesses derivatives of all orders at every point of the region and the function can be expanded in a power series about any interior point of the region.

By following Cauchy's development of complex variable theory, everything depends upon the complex integral calculus, and, in order to prove that a regular function possesses a second derivative, we must first of all express $f(z)$ as a contour integral round any closed contour surrounding the point z.

In order to develop the subject further we must now consider the definition of the integral of a function of a complex variable along a plane curve.

The equations $x = \phi(t)$, $y = \psi(t)$, where $a \leqslant t \leqslant \beta$, define the arc of a plane curve. If we subdivide the interval

(a, β) by the points $a = t_0, t_1, t_2, ..., t_r, ..., t_n = \beta$, then the points on the curve corresponding to these values of t may be denoted by $P_0, P_1, P_2, ..., P_n$. The length of the polygonal line $P_0P_1...P_n$, measured by $\sum_{r=1}^{n}\{(x_r - x_{r-1})^2 + (y_r - y_{r-1})^2\}^{\frac{1}{2}}$, depends on the particular mode of subdivision of (a, β). We call this summation the *length of an inscribed polygon*. If the arc be such that the lengths of all the inscribed polygons have a finite upper bound λ, the curve is said to be **rectifiable** and λ is **the length** of the curve.

It can be shown that the necessary and sufficient condition that the arc should be rectifiable is that the functions $\phi(t)$, $\psi(t)$ should be of *bounded variation* in (a, β). If $\phi'(t)$ and $\psi'(t)$ are *continuous*, it can be proved that the curve defined by $x = \phi(t)$, $y = \psi(t)$, $a \leqslant t \leqslant \beta$, is rectifiable and that its length s is given by *

$$s = \int_a^\beta \sqrt{[\{\phi'(t)\}^2 + \{\psi'(t)\}^2]}dt.$$

If we consider an arc of a Jordan curve whose equation is $z = \phi(t) + i\psi(t)$, where $a \leqslant t \leqslant \beta$, we define a **regular arc** of a Jordan curve to be one for which $\phi'(t)$, $\psi'(t)$ are continuous in $a \leqslant t \leqslant \beta$. From the above theorems we see that the length of this regular Jordan arc is

$$\int_a^\beta \sqrt{[\{\phi'(t)\}^2 + \{\psi'(t)\}^2]}dt.$$

By **a contour** we mean a continuous Jordan curve consisting of a finite number of regular arcs. Clearly a contour is rectifiable.

We now define the integral of a function of a complex variable z along a regular arc L defined by $x = \phi(t)$, $y = \psi(t)$, $a \leqslant t \leqslant \beta$.

* For proofs and further details, see P.A., p. 205 *seq.*, or G.I., p. 113.

Let $f(z)$ be any complex function of z, continuous along L, a regular arc with end-points A and B, and write $f(z) = u(x, y) + iv(x, y)$. Let $z_0, z_1, \ldots, z_n$ be points on L, z_0 being A and z_n being B. Consider the sum

$$\sum_{r=1}^{n} \{f(\zeta_r)(z_r - z_{r-1})\}, \qquad \cdot \qquad \cdot \qquad \cdot \quad (1)$$

where ζ_r is any point in the arc z_{r-1}, z_r. If $\zeta_r = \xi_r + i\eta_r$, we write $u_r = u(\xi_r, \eta_r)$, $v_r = v(\xi_r, \eta_r)$, and (1) may be written

$$\sum_{r=1}^{n} \{(u_r + iv_r)(x_r + iy_r - x_{r-1} - iy_{r-1})\}.$$

Now, by the mean-value theorem,

$$x_r - x_{r-1} = \phi(t_r) - \phi(t_{r-1}) = (t_r - t_{r-1})\phi'(\tau_r),$$
$$y_r - y_{r-1} = \psi(t_r) - \psi(t_{r-1}) = (t_r - t_{r-1})\psi'(\tau_r'),$$

where $t_{r-1} \leqslant \tau_r \leqslant t_r$, $t_{r-1} \leqslant \tau_r' \leqslant t_r$. Hence the sum may be written

$$\sum_{r=1}^{n} [(u_r + iv_r)\{\phi'(\tau_r) + i\psi'(\tau_r')\}(t_r - t_{r-1})]. \qquad \cdot \quad (2)$$

Since all the functions concerned are continuous, and therefore *uniformly continuous*, we can, given ϵ, find $\delta(\epsilon)$ so that

$$|u_r\phi'(\tau_r) - u(x_r, y_r)\phi'(t_r)| < \epsilon$$

for every r, provided that each $|t_r - t_{r-1}| < \delta$. Also

$$\sum_{r=1}^{n} \{\epsilon(t_r - t_{r-1})\} = \epsilon(t_n - t_0).$$

It follows that, as ϵ and δ tend to zero,

$$\sum_{r=1}^{n} \{u_r\phi'(\tau_r)(t_r - t_{r-1})\}$$

tends to the same limit as

$$\sum_{r=1}^{n} \{u(x_r, y_r)\phi'(t_r)(t_r - t_{r-1})\},$$

that is to the limit

$$\int_{\alpha}^{\beta} u\{\phi(t),\ \psi(t)\}\phi'(t)dt.$$

Similarly the other terms of (2) tend to limits, and we find that the whole sum tends to the limit

$$\int_{\alpha}^{\beta} [(u+iv)\{\phi'(t)+i\psi'(t)\}]dt. \qquad . \qquad . \quad (3)$$

This limit (3) is taken as the definition of the complex integral of $f(z)$ along the regular arc L, and it is written

$$\int_{L} f(z)dz.$$

The integral of $f(z)$ along a contour C, consisting of a finite number of regular arcs L_r, is given by

$$\int_{C} f(z)dz = \underset{r}{\Sigma} \int_{L_r} f(z)dz.$$

§ 31. An Upper Bound for a Contour Integral

I. *If $f(z)$ is continuous on a contour L, of length l, on which it satisfies the inequality $|f(z)| \leqslant M$, then*

$$\left| \int_{L} f(z)dz \right| \leqslant Ml.$$

It suffices to prove this theorem for a regular arc L. Since the modulus of any integral of a function of a real variable cannot exceed the integral of the modulus of that function, we have

$$\left| \int_{L} f(z)dz \right| = \left| \int_{\alpha}^{\beta} [(u+iv)\{\phi'(t)+i\psi'(t)\}]dt \right|,$$

$$\leqslant \int_{\alpha}^{\beta} M[\{\phi'(t)\}^2+\{\psi'(t)\}^2]^{\frac{1}{2}}dt,$$

$$= Ml.$$

If C is a *closed contour* we make the convention that the *positive* sense of description of the contour is *anti-clockwise*.

§ 32. Cauchy's Theorem

The elementary proof of Cauchy's theorem, which depends on the two-dimensional form of Green's theorem, requires the assumption of the *continuity* of $f'(z)$. We first give a proof with this assumption, but, on account of the fundamental importance of Cauchy's theorem in complex variable theory, we shall also prove the theorem under less restrictive assumptions.

II. *The elementary proof of Cauchy's theorem.*

If $f(z)$ is a regular function and if $f'(z)$ is continuous at each point within and on a closed contour C, then

$$\int_C f(z)dz = 0. \qquad . \qquad . \qquad . \quad (1)$$

Let D be the closed domain which consists of all points within and on C. Then by § 30 (3) we can write the integral (1) as a combination of curvilinear integrals

$$\int_C f(z)dz = \int_C (udx-vdy)+i\int_C (vdx+udy).$$

We transform each of these integrals by Green's theorem,* which states that, if $P(x, y)$, $Q(x, y)$, $\partial Q/\partial x$, $\partial P/\partial y$ are all continuous functions of x and y in D, then

$$\int_C (Pdx+Qdy) = \iint_D \left(\frac{\partial Q}{\partial x}-\frac{\partial P}{\partial y}\right) dxdy.$$

Since $f'(z) = u_x+iv_x = v_y-iu_y$, and, by hypothesis, $f'(z)$ is continuous in D, the conditions of Green's theorem are satisfied and so

$$\int_C f(z)dz = -\iint_D \left(\frac{\partial v}{\partial x}+\frac{\partial u}{\partial y}\right) dxdy+i\iint_D \left(\frac{\partial u}{\partial x}-\frac{\partial v}{\partial y}\right)dxdy$$
$$= 0,$$

by virtue of the Cauchy-Riemann equations.

* See P.A., pp. 290-1, or G.I., p. 54.

It was first shown by Goursat that it is unnecessary to assume the *continuity* of $f'(z)$ and that Cauchy's theorem holds if we only assume that $f'(z)$ *exists* at all points within and on C. In fact the continuity of $f'(z)$, and indeed its differentiability, are consequences of Cauchy's theorem.

Second proof of Cauchy's theorem.

If $f(z)$ is regular at all points within and on the closed contour C then

$$\int_C f(z)dz = 0.$$

The integral certainly exists, for a regular function $f(z)$ is continuous and a continuous function is integrable. We observe also that, if we construct a network of squares, by lines parallel to the axes of x and y, having the contour C as outer boundary, then C is divided into a network of meshes, either squares or parts of squares, such that

$$\int_C f(z)dz = \Sigma \int_\gamma f(z)dz,$$

where γ denotes the boundary of a mesh described in the same sense as C.

If z_0 lies inside a square contour S of side a, then

$$\left| \int_S | z-z_0 | \ | dz | \ \right| < 4\sqrt{2}a^2 = 4\sqrt{2}(\text{Area of } S).$$

This follows at once from § 31, for $| z-z_0 | < a\sqrt{2}$ and the length of the contour S is $4a$.

We now prove two lemmas.

Lemma 1.   *If C is a closed contour,* $\int_C dz = 0, \int_C zdz = 0.$

These results both follow from the definition of the integral, for

$$\int_C dz = \lim \sum_{1}^{n} \{(z_r - z_{r-1}).1\} = 0, \text{ as max } | z_r - z_{r-1} | \to 0.$$

Also

$$\int_C z\,dz = \lim \Sigma\{z_r(z_r-z_{r-1})\} = \lim \Sigma\{z_{r-1}(z_r-z_{r-1})\}$$
$$= \tfrac{1}{2} \lim \Sigma\{(z_r+z_{r-1})(z_r-z_{r-1})\} = \tfrac{1}{2} \lim \Sigma\{z_r^2-z_{r-1}^2\}$$
$$= 0.$$

Lemma 2. *Goursat's Lemma. Given ϵ, then, by suitable transversals, we can divide the interior of C into a finite number of meshes, either complete squares or parts of squares, such that, within each mesh, there is a point z_0 such that*

$$f(z)-f(z_0) = f'(z_0)(z-z_0)+\epsilon_\gamma\,|\,z-z_0\,|, \quad \bullet \qquad \bullet \quad (1)$$

for all values of z in the mesh, where $|\,\epsilon_\gamma\,|<\epsilon$.

Suppose the lemma is false ; then, however the interior of C is subdivided, there will be at least one mesh for which (1) is untrue. We shall show that this necessarily implies the existence of a point within or on C at which $f(z)$ is not differentiable.

Enclose C in a large square Γ, of area A, and apply the process of repeated quadrisection. When Γ is quadrisected there is at least one of the four quarters of Γ for which (1) does not hold. Let Γ_1 be the one chosen. Quadrisect Γ_1, choose one quarter of Γ_1, and so on. We thus obtain an unending sequence of squares Γ_1, Γ_2, ..., Γ_n, ..., each contained in the preceding, for which the lemma is untrue. These squares determine a limit-point ζ, and it is clear that ζ must lie within C.

Since $f(z)$ is differentiable at ζ,

$$f(z)-f(\zeta) = f'(\zeta)(z-\zeta)+\epsilon_\zeta\,|\,z-\zeta\,|,$$

where, for sufficiently small values of $|\,z-\zeta\,|$, $|\epsilon_\zeta|<\epsilon$. Now all the Γ_r, from one particular one onwards, lie within a circle, of centre ζ, for which $|\,z-\zeta\,|$ is so small that $|\,\epsilon_\zeta\,|<\epsilon$. This gives a contradiction, for by taking ζ to be z_0, (1) is satisfied. This proves Goursat's lemma.

Proof of the theorem. Some of the meshes γ obtained by the subdivision of the interior of C will be squares, others will be irregular, since we are not concerned with the exterior of C.

Integrate (1) round the boundary of each mesh. By virtue of lemma 1, we get

$$\int_\gamma f(z)dz = \int_\gamma \epsilon_\gamma \mid z-z_0 \mid dz \;;$$

and so, by addition,

$$\int_C f(z)dz = \Sigma \int_\gamma \epsilon_\gamma \mid z-z_0 \mid dz,$$

where

$$\left| \int_C f(z)dz \right| \leqslant \Sigma \left| \int_\gamma \epsilon_\gamma \mid z-z_0 \mid \mid dz \mid \right|.$$

If γ is not a complete square, divide it into two parts, γ_1 consisting of straight pieces, γ_2 consisting of parts of C. Since $\mid \epsilon_\zeta \mid < \epsilon$,

$$\Sigma \left| \int_{\gamma_1} \epsilon_\gamma \mid z-z_0 \mid \mid dz \mid \right| < \epsilon \cdot 4\sqrt{2}A,$$

A being the area of the large square Γ surrounding C. Also, the sum of the lengths of the portions γ_2 cannot exceed the length l of the rectifiable curve C, and so

$$\Sigma \left| \int_{\gamma_2} \epsilon_\gamma \mid z-z_0 \mid \mid dz \mid \right| < Kl\epsilon,$$

where K is the length of the diagonal of Γ, since $\mid z-z_0 \mid \leqslant K$. We deduce that

$$\left| \int_C f(z)dz \right| < B\epsilon,$$

where B is a constant, and, since ϵ is arbitrary, the theorem is proved.

§ 33. Cauchy's Integral, and the Derivatives of a Regular Function

By means of Cauchy's integral we can express the value of a regular function $f(z)$ at any point within a closed contour C as a contour integral round C.

III. *If $f(z)$ is regular within and on a closed contour C, and if ζ be a point within C, then*

$$\frac{1}{2\pi i}\int_C \frac{f(z)dz}{z-\zeta} = f(\zeta).$$

Describe about $z = \zeta$ a small circle γ of radius δ lying entirely within C. In the region between C and γ the

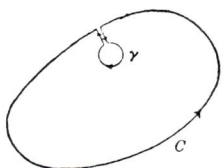

Fɪɢ. 12.

function $\phi(z) = f(z)/(z-\zeta)$ is regular. By making a cross-cut joining any point of γ to any point of C we form a closed contour Γ within which $\phi(z)$ is regular, so that, by Cauchy's theorem,

$$\int_\Gamma \phi(z)dz = 0.$$

In traversing the contour Γ in the positive (counter-clockwise) sense, the cross-cut is traversed twice, once in each sense, and so it follows that

$$\int_C \phi(z)dz - \int_\gamma \phi(z)dz = 0.$$

Now

$$\frac{1}{2\pi i}\int_\gamma \phi(z)dz = \frac{1}{2\pi i}\int_\gamma \frac{f(z)dz}{z-\zeta} = \frac{1}{2\pi i}\int_\gamma \frac{f(\zeta)dz}{z-\zeta} + \frac{1}{2\pi i}\int_\gamma \frac{f(z)-f(\zeta)}{z-\zeta}\,dz \quad . \quad (2)$$

Now on γ, $z - \zeta = \delta e^{i\theta}$, and so the first of the two terms on the right becomes

$$\frac{f(\zeta)}{2\pi i} \int_0^{2\pi} \frac{\delta i e^{i\theta} d\theta}{\delta e^{i\theta}} = f(\zeta)$$

and, by theorem I, the modulus of second term on the right of (2) cannot exceed

$$\frac{1}{2\pi\delta} \max_\gamma |f(z) - f(\zeta)| \cdot 2\pi\delta.$$

Since $f(z)$ is continuous at $z = \zeta$ this expression tends to zero as $\delta \to 0$: this proves the theorem.

The next theorem shows how to find the value of $f'(\zeta)$ as a contour integral.

IV. *If $f(z)$ is regular in a domain D, its derivative is given by*

$$f'(\zeta) = \frac{1}{2\pi i} \int_C \frac{f(z) dz}{(z - \zeta)^2},$$

where C is any simple closed contour in D surrounding the point $z = \zeta$.

We have, by III,

$$\frac{f(\zeta + h) - f(\zeta)}{h} = \frac{1}{2\pi i} \int_C \frac{1}{h} \left(\frac{1}{z - \zeta - h} - \frac{1}{z - \zeta} \right) f(z) dz,$$

$$= \frac{1}{2\pi i} \int_C \left\{ \frac{f(z)}{(z - \zeta)^2} + \frac{h f(z)}{(z - \zeta)^2 (z - \zeta - h)} \right\} dz,$$

$$= \frac{1}{2\pi i} \int_C \frac{f(z) dz}{(z - \zeta)^2} + I.$$

If we now prove that $|I| \to 0$ as $|h| \to 0$, the required result is established. Since $f(z)$ is regular in and on C it is bounded, so that $|f(z)| \leqslant M$ on C. Let d be the lower bound of the distance of ζ from C : suppose h chosen so small that $| h |<\frac{1}{2}d$, then

$$| I | < \frac{| h |}{2\pi} \frac{Ml}{d^2 \cdot \frac{1}{2}d},$$

where l is the length of C. It is now clear that the term on the right tends to zero as $| h |\rightarrow 0$.

V. *If $f(z)$ is regular in a domain D, then $f(z)$ has, at every point ζ of D, derivatives of all orders, their values being given by*

$$f^{(n)}(\zeta) = \frac{n!}{2\pi i}\int_C \frac{f(z)dz}{(z-\zeta)^{n+1}}.$$

If we assume the theorem proved for $n = m$ and consider the expression

$$\frac{f^{(m)}(\zeta+h)-f^{(m)}(\zeta)}{h} ;$$

we can readily prove that it is equal to

$$\frac{(m+1)!}{2\pi i} \int_C \frac{f(z)dz}{(z-\zeta)^{m+2}} + I,$$

and the proof that $| I |$ tends to zero as $| h |\rightarrow 0$ follows the same lines as in IV. The details are left to the reader.

§ 34. Taylor's Theorem

VI. *If $f(z)$ is regular in $| z-a |\leqslant\rho$, and if ζ is a point such that $| \zeta-a | = r(<\rho)$ then*

$$f(\zeta) = \sum_{n=0}^{\infty} a_n(\zeta-a)^n,$$

where $a_n = f^{(n)}(a)/n!$.

Let C be a circle of radius ρ', centre $z = a$, where $r<\rho'<\rho$, and consider the identity

$$\frac{1}{z-\zeta} \equiv \frac{1}{z-a} + \frac{\zeta-a}{(z-a)^2} + \dots + \frac{(\zeta-a)^{n-1}}{(z-a)^n} + \frac{(\zeta-a)^n}{(z-a)^n} \frac{1}{z-\zeta}.$$

Multiply each term by $f(z)/2\pi i$ and integrate round C; we clearly obtain

$$f(\zeta) = f(a)+f'(a)(\zeta-a) + \dots + \frac{f^{(n-1)}(a)}{(n-1)!}\,(\zeta-a)^{n-1}+R_n,$$

where

$$R_n = \frac{(\zeta-a)^n}{2\pi i}\int_C \frac{f(z)dz}{(z-a)^n(z-\zeta)}\,.$$

This is Taylor's theorem with remainder.

Since $|f(z)|\leqslant M$ on C we readily see that

$$|R_n| \leqslant \frac{r^n}{2\pi}\frac{2\pi\rho' M}{\rho'^n(\rho'-r)} = K\left(\frac{r}{\rho'}\right)^n,$$

where K is a constant independent of n. Since $r<\rho'$ we see that $|R_n|\to 0$ as $n\to\infty$.

It therefore follows that the series $\overset{\infty}{\underset{0}{\Sigma}}a_n(\zeta-a)^n$ is convergent and has $f(\zeta)$ as its sum-function. If $f(z)$ is regular in the whole z-plane, the expansion is valid for all ζ.

Corollary. If $|f(\zeta)|$ has a maximum $M(r)$ on $|\zeta-a| = r<\rho$ then, if $a_n = f^{(n)}(a)/n!$, we have the inequality

$$|a_n| \leqslant \frac{M(r)}{r^n}\,.$$

For, if C be the circle $|z-a| = r$ we have

$$|a_n| = \left|\frac{1}{2\pi i}\int_C \frac{f(z)dz}{(z-a)^{n+1}}\right| \leqslant \frac{1}{2\pi}\frac{M(r)}{r^{n+1}}\,2\pi r = \frac{M(r)}{r^n}\,.$$

§ 35. The Theorems of Liouville and Laurent

VII. Liouville's Theorem. *If $f(z)$ is regular in the whole z-plane and if $|f(z)|<K$ for all values of z, then $f(z)$ must be a constant.*

Let z_1, z_2 be any two points and C a circle of centre z_1

and the radius $\rho \geqslant 2 \mid z_1 - z_2 \mid$, so that, when z is on C, $\mid z - z_2 \mid \geqslant \frac{1}{2}\rho$. By III,

$$f(z_1) - f(z_2) = \frac{1}{2\pi i} \int_C \left(\frac{1}{z - z_1} - \frac{1}{z - z_2} \right) f(z)dz,$$

so that

$$\mid f(z_1) - f(z_2) \mid = \frac{1}{2\pi} \left| \int_C \frac{(z_1 - z_2)f(z)dz}{(z - z_1)(z - z_2)} \right| < \frac{1}{2\pi} \int_0^{2\pi} \frac{\mid z_1 - z_2 \mid K}{\frac{1}{2}\rho} d\theta$$

$$= 2 \mid z_1 - z_2 \mid \frac{K}{\rho}.$$

Keep z_1 and z_2 fixed and make $\rho \to \infty$, then it follows that $f(z_1) = f(z_2)$; in other words, $f(z)$ is a constant.

VIII. Laurent's Theorem. *Let C_1 and C_2 be two circles of centre a with radii ρ_1 and ρ_2 $(\rho_2 < \rho_1)$; then, if $f(z)$ be regular on the circles and within the annulus between C_1 and C_2,*

$$f(\zeta) = \overset{\infty}{\underset{0}{\Sigma}} a_n(\zeta - a)^n + \overset{\infty}{\underset{1}{\Sigma}} b_n(\zeta - a)^{-n}$$

ζ being any point of the annulus. The coefficients a_n and b_n are given by

$$a_n = \frac{1}{2\pi i} \int_{C_1} \frac{f(z)dz}{(z - a)^{n+1}}, \quad b_n = \frac{1}{2\pi i} \int_{C_2} (z - a)^{n-1} f(z)dz.$$

By making a cross-cut joining any point of C_1 to any point of C_2, we readily see that

$$f(\zeta) = \frac{1}{2\pi i} \int_{C_1} \frac{f(z)dz}{z - \zeta} - \frac{1}{2\pi i} \int_{C_2} \frac{f(z)dz}{z - \zeta}.$$

Consider the two identities

$$\frac{1}{z - \zeta} \equiv \frac{1}{z - a} + \frac{\zeta - a}{(z - a)^2} + \ldots + \frac{(\zeta - a)^{n-1}}{(z - a)^n} + \frac{(\zeta - a)^n}{(z - a)^n} \frac{1}{z - \zeta},$$

$$-\frac{1}{z - \zeta} \equiv \frac{1}{\zeta - a} + \frac{z - a}{(\zeta - a)^2} + \ldots + \frac{(z - a)^{n-1}}{(\zeta - a)^n} + \frac{(z - a)^n}{(\zeta - a)^n} \frac{1}{\zeta - z}.$$

Then it follows that

$$\frac{1}{2\pi i}\int_{C_1}\frac{f(z)dz}{z-\zeta}=\sum_{r=0}^{n-1}a_r(\zeta-a)^r+P_n,$$

$$-\frac{1}{2\pi i}\int_{C_2}\frac{f(z)dz}{z-\zeta}=\sum_{r=1}^{n}b_r(\zeta-a)^{-r}+Q_n,$$

where

$$P_n=\frac{(\zeta-a)^n}{2\pi i}\int_{C_1}\frac{f(z)dz}{(z-a)^n(z-\zeta)},\ Q_n=\frac{(\zeta-a)^{-n}}{2\pi i}\int_{C_2}\frac{f(z)dz}{(z-a)^{-n}(\zeta-z)}$$

Now P_n is precisely the same remainder as in Taylor's theorem and we can prove, in the same way as in VI, that $|P_n|\to 0$ as $n\to\infty$.

Also we have

$$|Q_n|=\frac{1}{2\pi}\left|\int_{C_2}\left(\frac{z-a}{\zeta-a}\right)^n\frac{f(z)dz}{z-\zeta}\right|\leqslant\frac{1}{2\pi}\left(\frac{\rho_2}{r}\right)^n\frac{M'}{r-\rho_2}2\pi\rho_2,$$

where $r=|\zeta-a|$ and $|f(z)|\leqslant M'$ on C_2. Since $\rho_2<r$, it follows that $|Q_n|\to 0$ as $n\to\infty$.

If, therefore, we write

$$f(\zeta)=f_1(\zeta)+f_2(\zeta),$$

where $f_1(\zeta)=\sum_0^\infty a_n(\zeta-a)^n$ and $f_2(\zeta)=\sum_0^\infty b_n(\zeta-a)^{-n}$, we see that $f(\zeta)$ converges for $\rho_2\leqslant|\zeta-a|\leqslant\rho_1$.

It also follows that $f_1(\zeta)$ is regular and converges for $|\zeta-a|\leqslant\rho_1$ and that $f_2(\zeta)$ is regular and converges for $|\zeta-a|\geqslant\rho_2$.

§ 36. Zeros and Singularities

If $f(z)$ is regular within a given domain D, we have seen that it can be expanded in a Taylor series about any point $z=a$ of D and

$$f(z)=\sum_0^\infty a_n(z-a)^n.$$

If $a_0 = a_1 = \ldots = a_{m-1} = 0$, $a_m \neq 0$, the first term in the Taylor expansion is $a_m(z-a)^m$. In this case $f(z)$ is said to have a **zero** of order m at $z = a$.

A **singularity** of a function $f(z)$ is a point at which the function ceases to be regular.

If $f(z)$ is regular within a domain D, *except at the point $z = a$*, which is an isolated singularity of $f(z)$, then we can draw two concentric circles of centre a, both lying within D. The radius of the smaller circle ρ_2 may be as small as we please, and the radius ρ_1 of the larger circle of any length, subject to the restriction that the circle lies wholly within D. In the annulus between these two circles, $f(z)$ has a Laurent expansion of the form

$$f(z) = \sum_0^\infty a_n(z-a)^n + \sum_1^\infty b_n(z-a)^{-n}.$$

The second term on the right is called the **principal part** of $f(z)$ at $z = a$.

It may happen that $b_m \neq 0$ while $b_{m+1} = b_{m+2} \ldots = 0$. In this case the principal part consists of the finite number of terms

$$\frac{b_1}{z-a} + \frac{b_2}{(z-a)^2} + \ldots + \frac{b_m}{(z-a)^m},$$

and the singularity at $z = a$ is called a **pole** of **order** m of $f(z)$ and the coefficient b_1, which may in certain cases be zero, is called the **residue** of $f(z)$ at the pole $z = a$. If the pole be of order *one*, $b_1 = \lim_{z \to a}\{(z-a)f(z)\}$.

If the principal part is an infinite series, the singularity is an *isolated essential singularity*.

(1) If $z = a$ is a zero of order m of $f(z)$, we now prove that this *zero is isolated*: in other words, *there exists a neighbourhood of the point $z = a$ which contains no other zero of $f(z)$.*

Clearly we can write $f(z) = (z-a)^m\phi(z)$, where $\phi(z)$ is regular in $|z-a| < \rho$ and $\phi(a) \neq 0$, since $\phi(a) = a_m$.

Write $\phi(a) = 2c$, then, since $\phi(z)$ is continuous, there exists a region $|z-a| < \delta$ in which $|\phi(z)-\phi(a)| < |c|$. Hence

$$|\phi(z)| \geqslant |\{|\phi(a)| - |\phi(z)-\phi(a)|\}| > |c|,$$

where $|z-a| < \delta$, and so $\phi(z)$ does not vanish in $|z-a| < \delta$.

(2) If $z = a$ is a pole of order m of $f(z)$ it follows, from the definition of a pole by means of Laurent's theorem, that *poles are isolated*, for, the small circle, of centre $z = a$ and radius ρ_2, encloses the *only* singularity of $f(z)$ within the domain D which contains the annulus between the two circles of radii ρ_1 and ρ_2.

(3) *If $f(z)$ has a pole at $z = a$, then $|f(z)| \to \infty$ as $z \to a$ in any manner.* For, if the pole be of order m,

$$f(z) = (z-a)^{-m}\{b_m + b_{m-1}(z-a) + \ldots + b_1(z-a)^{m-1} + \sum_0^\infty a_n(z-a)^{n+m}\}$$

and, since $b_m \neq 0$, we may write $f(z) = (z-a)^{-m}\psi(z)$, where $\psi(z)$ is regular in $|z-a| < \rho$, and $\psi(a) = b_m (\neq 0)$. Hence, by (1), we can find a neighbourhood $|z-a| < \delta$ of the pole in which $|\psi(z)| > \frac{1}{2}|b_m|$, from which it follows that

$$|f(z)| > \tfrac{1}{2}|b_m||z-a|^{-m}.$$

Hence $|f(z)| \to \infty$ as $z \to a$ in any manner.

(4) *Limit points of zeros and poles.*

Let $a_1, a_2, \ldots, a_n, \ldots$ be a sequence of zeros of a function $f(z)$ which is regular in a domain D. Suppose that these zeros have a limit point α which is an interior point of D. Since $f(z)$ is a continuous function, having zeros as near as we please to α, $f(\alpha)$ must be zero. Now $z = \alpha$ cannot be a zero of $f(z)$, since we have proved in (1) that zeros are isolated. Hence $f(z)$ must be identically zero.

If $f(z)$ is not identically zero in D, then $z = \alpha$ must be a singularity of $f(z)$. The singularity is isolated, but it is not a pole, since $|f(z)|$ does not tend to infinity as $z \to \alpha$ in any manner. Hence a limit point of zeros must be an **isolated essential singularity** of $f(z)$.

If $f(z)$ be regular, except at a set of points which are singularities c_1, c_2, ..., c_n, ..., infinite in number, and having a limit point γ in D, then γ must be a singularity of $f(z)$, since $f(z)$ is unbounded in the neighbourhood of γ. Since γ is not isolated it cannot be a pole. We call such a singularity a **non-isolated essential singularity**.

Examples. sin $1/z$ has an isolated essential singularity at $z = 0$. It is the limit point of the zeros, $z = 1/n\pi$, $(n = \pm 1, \pm 2, ...)$. tan $1/z$ has poles at the points $z = 2/n\pi$, $(n = \pm 1, \pm 3, ...)$, and so the limit point of the poles, $z = 0$, is a non-isolated essential singularity.

Note on the region of convergence of a Taylor series.

If $f(z)$ be a function which is regular, except at a number of isolated singularities at finite points of the z-plane, then we can expand $f(z)$ in a Taylor series $\sum\limits_{0}^{\infty} a_n(z-a)^n$ about any assigned point $z = a$, and the radius of convergence ρ of this power series will be the distance from $z = a$ to the nearest singularity of $f(z)$, since $f(z)$ is clearly regular in $|z-a| < \rho$, and cannot be regular in any circle of centre a whose radius exceeds ρ.

We see that the radius of convergence of a power series depends upon the extent of the region within which the sum-function is regular, and it may be controlled by the existence of singularities which do not necessarily lie on the real axis.

If we consider the real function $1/(1-x)$, the binomial expansion leads to

$$\frac{1}{1-x} = 1+x+x^2+..., \qquad . \qquad . \quad (1)$$

the series being convergent if $|x| < 1$. This seems quite natural, since the sum-function has a singularity at $x = 1$. However, on considering the function $1/(1+x^2)$, we have

$$\frac{1}{1+x^2} = 1-x^2+x^4-... \qquad . \qquad . \quad (2)$$

and the series is again convergent only if $|x|<1$; but if we regard $1/(1+x^2)$ as a function of the *real* variable x there is nothing in the nature of the function to suggest the restriction of its range of convergence to $|x|<1$. If, however, we consider $1/(1+z^2)$, where z is complex, the restriction on the region of convergence is at once evident, since $1/(1+z^2)$ has singularities at $z=\pm i$, and the radius of convergence of the series (2), if x is complex, is the distance of the origin from the nearest singularity and this is plainly unity.

§ 37. The Point at Infinity

In complex variable theory we have seen that it is convenient to regard infinity as a single point. The behaviour of $f(z)$ " at infinity " is considered by making the substitution $z=1/\zeta$ and examining $f(1/\zeta)$ at $\zeta=0$. We say that $f(z)$ is regular, or has a simple pole, or has an essential singularity at infinity according as $f(1/\zeta)$ has the corresponding property at $\zeta=0$.

We know that if $f(1/\zeta)$ has a pole of order m at $\zeta=0$, near $\zeta=0$ we have

$$f(1/\zeta)=\sum_{n=0}^{\infty} a_n\zeta^n + \frac{b_1}{\zeta} + \frac{b_2}{\zeta^2} + ... + \frac{b_m}{\zeta^m},$$

and so, near $z=\infty$,

$$f(z)=\sum_{n=0}^{\infty} a_n z^{-n}+b_1 z+b_2 z^2+...+b_m z^m.$$

Thus, when $f(z)$ has a pole of order m at infinity, the principal part of $f(z)$ at infinity is the finite series in ascending powers of z.

Since

$$\sin z = z - \frac{z^3}{3!} + \frac{z^5}{5!} - ...,$$

the function $\sin z$ has an isolated essential singularity at infinity, the principal part at infinity being an infinite series.

§ 38. Rational Functions

Theorem. If a single-valued function $f(z)$ has no essential singularities either in the finite part of the plane or at infinity, then $f(z)$ is a rational function.

Since the point at infinity is not an essential singularity of $f(z)$, we can surround it by a region in which $f(z)$ either is regular or has the point at infinity as its *only* singularity. That is, we can draw a circle C, with centre the origin, such that the point at infinity is the only singularity outside C. There can only be a finite number of singularities within C, since poles are isolated singularities. Suppose that the poles inside C are at $a_1, a_2, \ldots, a_n$. The principal part at a_s may be written

$$\frac{b_{1s}}{z-a_s} + \frac{b_{2s}}{(z-a_s)^2} + \ldots + \frac{b_{ms}}{(z-a_s)^m},$$

a_s being supposed to be a pole of order m. The principal part at infinity is of the form

$$B_1 z + B_2 z^2 + \ldots + B_k z^k.$$

Now consider the function

$$\phi(z) = f(z) - \sum_{s=1}^{n} \left\{ \frac{b_{1s}}{z-a_s} + \ldots + \frac{b_{ms}}{(z-a_s)^m} \right\} - (B_1 z + \ldots + B_k z^k).$$

The function $\phi(z)$ is plainly regular everywhere in the plane, even at infinity: hence $\phi(z)$ is bounded for all z, and so, by Liouville's theorem, $\phi(z)$ is a constant. Hence

$$f(z) = C + \sum_{s=1}^{n} \left\{ \frac{b_{1s}}{z-a_s} + \ldots + \frac{b_{ms}}{(z-a_s)^m} \right\} + B_1 z + \ldots + B_k z^k,$$

and so $f(z)$ is a rational function of z.

§ 39. Analytic Continuation

Suppose that $f_1(z)$ and $f_2(z)$ are functions regular in domains D_1 and D_2 respectively and that D_1 and D_2 have a common part, throughout which $f_1(z) = f_2(z)$, then we regard the aggregate of values of $f_1(z)$ and $f_2(z)$, at points interior to D_1 or D_2 as a single regular function $\phi(z)$. Thus $\phi(z)$ is regular in $\Delta = D_1 + D_2$ and $\phi(z) = f_1(z)$ in D_1 and $\phi(z) = f_2(z)$ in D_2. The function $f_2(z)$ may be regarded as extending the domain in which $f_1(z)$ is defined and it is called an **analytic continuation** of $f_1(z)$.

The standard method of continuation is the method of power series which we now briefly describe.

Let P be the point z_0, in the neighbourhood of which $f(z)$ is regular, then, by Taylor's theorem, we can expand $f(z)$ in a series of ascending powers of $z-z_0$, the coefficients in which involve the successive derivatives of $f(z)$ at z_0. If S be the singularity of $f(z)$ which is nearest to P, then the Taylor expansion is valid within a circle of centre P and radius PS. Now choose any point P_1 within the circle of convergence not on the line PS. We can find the values of $f(z)$ and all its derivatives at P_1, from the series, by repeated term-by-term differentiation, and so we can form the Taylor series for $f(z)$ with P_1 as origin, and this series will define a function regular in some circle of centre P_1. This circle will extend as far as the singularity, of the function defined by the new series, which is nearest to P_1 and this may or may not be S. In either case the new circle of convergence may lie partly outside the old circle and, for points in the region included in the new circle but not in the old, the new series may be used to define the values of $f(z)$ although the old series failed to do so.

Similarly, we may take any other point P_2 in the region for which the values of the function are now known and form the Taylor series with P_2 as origin which will, in general, still further extend the region of definition of the function ; and so on.

By means of this process of continuation, starting from a representation of a function by any one power series, we can find any number of other power series, which between them define the value of the function at all points of a domain, any point of which can be reached from P without passing through a singularity of the function. It can be proved that continuation by two different paths PQR, $PQ'R$ gives the same final power series provided that the function has no singularity inside the closed curve $PQRQ'P$.

We may now, following Weierstrass, define an **analytic function** of z as one power series together with all the other power series which can be derived from it by analytic continuation. Two different analytic expressions then define the same function if they represent power series derivable from each other by continuation. The complete analytic function defined in this way need not be a one-valued function. Each of the continuations is called an **element** of the analytic function.

If $f(z)$ is not an integral function there will be certain exceptional points which do not lie in any of the domains into which $f(z)$ has been continued. These points are the *singularities* of the analytic function. Clearly the singular points of a one-valued function are also singularities in this wider sense.

There must be at least one singularity of the analytic function on the circle of convergence C_0 of the power series $\sum\limits_{0}^{\infty} a_n(z-z_0)^n$.* For, if not, we could construct, by continuation, a function equal to $f(z)$ within C_0 but regular in a larger concentric circle Γ_0. The expansion of this function in a Taylor series in powers of $z-z_0$ would then converge everywhere within Γ_0. This is impossible, since the series would be the original series whose circle of convergence is C_0. If z_1 is any point within C_0, let C_1 be the circle of convergence of the power series

$$\sum\limits_{0}^{\infty} f^{(n)}(z_1) \frac{(z-z_1)^n}{n!}.$$

* For a proof of this, and further details, see Titchmarsh, *Theory of Functions* (Oxford, 1932), p. 145.

If Q_1 is the circle of centre z_1 which touches C_0 internally, the new power series is certainly convergent within Q_1 and has the sum $f(z)$ there. There are now three possibilities. Since the radius of C_1 cannot be less than that of Q_1, we have either (i) C_1 has a larger radius than Q_1, or (ii) C_0 is a natural boundary * of $f(z)$, or (iii) C_1 may touch C_0 internally, though C_0 is not a natural boundary of $f(z)$.

In case (i) C_1 lies partly outside C_0 and the new power series provides an analytic continuation of $f(z)$: we can then take a point z_2 within C_1 and outside C_0 and repeat the process. In case (ii) we cannot continue $f(z)$ outside C_0 and the circle C_1 touches C_0 internally no matter what point z_1 within C_0 is chosen. In case (iii) the point of contact of C_0 and C_1 is a singularity of the analytic function obtained by continuation of the original power series. For there is necessarily one singularity on C_1 and this cannot be within C_0.

We may illustrate some of the above remarks by the following examples.

1. The series

$$\frac{1}{a} + \frac{z}{a^2} + \frac{z^2}{a^3} + \frac{z^3}{a^4} + \cdots$$

represents the function $f(z) = 1/(a-z)$ only for points within the circle $|z| = |a|$. If b/a is not real, the series

$$\frac{1}{a-b} + \frac{z-b}{(a-b)^2} + \frac{(z-b)^2}{(a-b)^3} + \cdots,$$

for different values of b, represents $f(z)$ at points outside the circle $|z| = |a|$.

2. That there are functions to which the process of continuation cannot be applied may be seen by considering the function

$$g(z) = 1 + z^2 + z^4 + \cdots + z^{2^n} + \cdots.$$

---

* See Example 2 below.

It is readily shown that any root of any of the equations

$$z^2 = 1, \; z^4 = 1, \; z^8 = 1, \; z^{16} = 1, \; \ldots,$$

is a singularity of $g(z)$, and hence that on any arc, however small, of the circle $|z| = 1$ there is an unlimited number of them. The circle $|z| = 1$ is in this case a *natural boundary* of $g(z)$. This illustrates case (ii) above.

§ 40. Poles and Zeros of Meromorphic Functions

A function $f(z)$, whose only singularities in the finite part of the plane are poles, is called a **meromorphic function**. We now prove a very useful theorem.

If $f(z)$ is meromorphic inside a closed contour C, and is not zero at any point on the contour, then

$$\frac{1}{2\pi i} \int_C \frac{f'(z)}{f(z)} \, dz = N - P, \qquad \bullet \qquad \bullet \qquad (1)$$

where N is the number of zeros and P the number of poles inside C. (A pole or zero of order m must be counted m times.)

Suppose that $z = a$ is a zero of order m, then, in the neighbourhood of this point

$$f(z) = (z-a)^m \phi(z),$$

where $\phi(z)$ is regular and not zero. Hence

$$\frac{f'(z)}{f(z)} = \frac{m}{z-a} + \frac{\phi'(z)}{\phi(z)}.$$

Since the last term is regular at $z = a$, we see that $f'(z)/f(z)$ has a simple pole at $z = a$ with residue m. Similarly, if $z = b$ is a pole of order k, we see that $f'(z)/f(z)$ has a simple pole at $z = b$ with residue $-k$. It follows, by § **33**, III, that the left-hand side of (1) is equal to $\Sigma m - \Sigma k = N - P$.

If $f(z)$ is regular in C, then $P = 0$, and the integral on the left of (1) is equal to N. Since

$$\frac{d}{dz} \log f(z) = \frac{f'(z)}{f(z)},$$

we may write the result in another form,

$$\int_C \frac{f'(z)}{f(z)}\, dz = \Delta_C \log f(z),$$

where Δ_C denotes the variation of $\log f(z)$ round the contour C. The value of the logarithm with which we start is immaterial ; and, since

$$\log f(z) = \log |f(z)| + i \arg f(z)$$

and $\log |f(z)|$ is one-valued, the formula may be written

$$N = \frac{1}{2\pi} \Delta_C \arg f(z).$$

This result is known as *the principle of the argument.*

§ 41. Rouché's Theorem

If $f(z)$ and $g(z)$ are regular within and on a closed contour C and $|g(z)| < |f(z)|$ on C, then $f(z)$ and $f(z)+g(z)$ have the same number of zeros inside C.

We observe that neither $f(z)$ nor $f(z)+g(z)$ has a zero on C, and so, if N is the number of zeros of $f(z)$ and N' the number of zeros of $f(z)+g(z)$,

$$2\pi N = \Delta_C \arg f,$$

$$2\pi N' = \Delta_C \arg (f+g) = \Delta_C \arg f + \Delta_C \arg \left(1 + \frac{g}{f}\right).$$

The theorem is proved if we show that

$$\Delta_C \arg \left(1 + \frac{g}{f}\right) = 0.$$

Since $|g| < |f|$, the point $w = 1+g/f$ is always an interior point of the circle of centre $w = 1$ and radius unity: thus, if $w = \rho e^{i\phi}$, ϕ always lies between $-\tfrac{1}{2}\pi$ and $\tfrac{1}{2}\pi$ and so $\arg (1+g/f) = \phi$ returns to its original value when z describes C. Since ϕ cannot increase or decrease by a multiple of 2π, the theorem follows.

The preceding theorems are useful for locating the roots of equations. The method is illustrated by the following example.

Example. Prove that one root of the equation $z^4+z^3+1 = 0$ lies in the first quadrant.

The equation $z^4+z^3+1 = 0$ plainly has no real roots. For, if we put $z = x$, $x^4+x^3+1 = 0$ has no real positive roots. If we put $z = -x$ and write $\phi(x) \equiv x^4-x^3+1 = 0$, we see that

$$\phi(x) = x^3(x-1)+1>0, \text{ if } x>1 ;$$
and $$\phi(x) = x^4+(1-x)(x^2+x+1)>0, \text{ if } 0<x<1.$$

Hence the given equation has no real negative roots.

The given equation has no purely imaginary roots either, for, on putting $z = iy$, we get $y^4-iy^3+1 = 0$ and it is plain that the real and imaginary parts never vanish together.

Consider $\Delta \arg (z^4+z^3+1)$ round part of the first quadrant bounded by $|z| = R$ where R is large. On the arc of the circle, $z = Re^{i\theta}$, and we have

$$\Delta \arg (z^4+z^3+1) = \Delta \arg (R^4e^{4i\theta})+\Delta \arg \{1+O(R^{-1})\},$$
$$= 2\pi+O(R^{-1}).$$

On the axis of y we have

$$\arg (z^4+z^3+1) = \arctan \left(\frac{-y^3}{y^4+1}\right)$$

The numerator of $-y^3/(y^4+1)$ only vanishes when $y = 0$ and the denominator does not vanish for any real y. Hence as y ranges from ∞ to 0 along the imaginary axis, the initial and final values of $\arg (z^4+z^3+1)$ are zero. Hence the total change in $\arg (z^4+z^3+1)$, where R is large, is 2π. It follows that one root of the given equation lies in the first quadrant.

§ 42. The Maximum–Modulus Principle

We now establish an important theorem which may be stated as follows.

If $f(z)$ is regular within and on a closed contour C, then $|f(z)|$ attains its maximum value on the boundary of C and not at any interior point.

Lemma.   If $\phi(x)$ is continuous, $\phi(x) \leqslant \kappa$ and

$$\frac{1}{b-a} \int_a^b \phi(x)dx \geqslant \kappa, \quad \bullet \qquad \bullet \qquad \bullet \quad (1)$$

then $\phi(x) = \kappa$.

Suppose that $\phi(x_1) < \kappa$, then there is an interval $(x_1-\delta, x_1+\delta)$ in which $\phi(x) \leqslant \kappa - \epsilon$ and

$$\int_a^b \phi(x)dx \leqslant 2\delta(\kappa-\epsilon) + (b-a-2\delta)\kappa = (b-a)\kappa - 2\delta\epsilon,$$

which contradicts (1).

Theorem.   If $|f(z)| \leqslant M$ on C, then $|f(z)| < M$ at all interior points of the domain D enclosed by C, unless $f(z)$ is a constant, in which case $|f(z)| = M$ everywhere.

Suppose that at an interior point z_0 of D, $|f(z)|$ has a value at least equal to its value elsewhere. Let Γ be a circle of centre z_0 lying entirely within D. Then by § 33, III,

$$f(z_0) = \frac{1}{2\pi i} \int_\Gamma \frac{f(z)dz}{z-z_0}. \quad \bullet \qquad \bullet \quad (2)$$

Write $z - z_0 = re^{i\theta}$, $f(z)/f(z_0) = \rho e^{i\phi}$, so that ρ and ϕ are functions of θ, then (2) may be written

$$1 = \frac{1}{2\pi} \int_0^{2\pi} \rho e^{i\phi} \, d\theta. \quad \bullet \qquad \bullet \quad (3)$$

Hence
$$1 \leqslant \frac{1}{2\pi} \int_0^{2\pi} \rho \, d\theta.$$

By hypothesis $\rho \leqslant 1$, and so, by the lemma, $\rho = 1$ for all values of θ.   On taking the real part of (3) we get

$$1 = \frac{1}{2\pi} \int_0^{2\pi} \cos\phi \, d\theta,$$

and so, by the lemma, $\cos\phi = 1$. Hence $f(z) = f(z_0)$ on Γ. Since $f(z)$ is a constant at any point a on Γ, it follows by Taylor's theorem that it is constant in a neighbourhood of a, and hence, by analytic continuation, $f(z)$ is constant everywhere within and on C.

There is a corresponding theorem for harmonic functions. *A function which is harmonic in a region cannot have a maximum at an interior point of the region.*

EXAMPLES IV

1. The function $f(z)$ is regular in $|z-a| < R$; prove that, if $0 < r < R$,

$$f'(a) = \frac{1}{\pi r} \int_0^{2\pi} P(\theta) e^{-i\theta} d\theta,$$

where $P(\theta)$ is the real part of $f(a + re^{i\theta})$.

2. $\phi(z)$ and $\psi(z)$ are two regular functions; $z = a$ is a once repeated root of $\psi(z) = 0$ and $\phi(a) \neq 0$. Prove that the residue of $\phi(z)/\psi(z)$ at $z = a$ is

$$\{6\,\phi'(a)\psi''(a) - 2\,\phi(a)\psi'''(a)\}/3\,\{\psi''(a)\}^2.$$

3. The function $f(z)$ is regular when $|z| < R'$. Prove that, if $|a| < R < R'$,

$$f(a) = \frac{1}{2\pi i} \int_C \frac{R^2 - a\bar{a}}{(z-a)(R^2 - z\bar{a})}\, f(z)dz,$$

where C is the circle $|z| = R$. Deduce Poisson's formula that, if $0 < r < R$,

$$f(re^{i\theta}) = \frac{1}{2\pi} \int_0^{2\pi} \frac{R^2 - r^2}{R^2 - 2Rr\cos(\theta - \phi) + r^2}\, f(Re^{i\phi})d\phi.$$

4. By using the integral representation of $f^{(n)}(a)$, (§ **33, V**), prove that

$$\left(\frac{x^n}{n!}\right)^2 = \frac{1}{2\pi i} \int_C \frac{x^n e^{xz}}{n!\, z^{n+1}}\, dz,$$

where C is any closed contour surrounding the origin. Hence prove that

$$\sum_{n=0}^{\infty} \left(\frac{x^n}{n!}\right)^2 = \frac{1}{2\pi} \int_0^{2\pi} e^{2x\cos\theta} d\theta.$$

5. Obtain the expansion

$$f(z) = f(a) + 2\left\{\frac{z-a}{2}f'\left(\frac{z+a}{2}\right) + \frac{(z-a)^3}{2^3 \cdot 3!}f'''\left(\frac{z+a}{2}\right) + \frac{(z-a)^5}{2^5 \cdot 5!}f^{(5)}\left(\frac{z+a}{2}\right) + \ldots\right\};$$

and determine its range of validity.

6. If $f(z) = \sum_{n=1}^{\infty} z^2/(4+n^2z^2)$, show that $f(z)$ is finite and continuous for all *real* values of z but $f(z)$ cannot be expanded in a Maclaurin series. Show that $f(z)$ possesses Laurent expansions valid in a succession of ring spaces.

7. Prove that $\cosh\left(z+\dfrac{1}{z}\right) = a_0 + \sum_{1}^{\infty} a_n\left(z^n+\dfrac{1}{z^n}\right)$, where

$$a_n = \frac{1}{2\pi}\int_0^{2\pi} \cos n\theta \cosh(2\cos\theta)d\theta.$$

8. Find the Taylor and Laurent series which represent the function $(z^2-1)/\{(z+2)(z+3)\}$ in (i) $|z|<2$, (ii) $2<|z|<3$, (iii) $|z|>3$.

9. Find the nature and location of the singularities of the function $1/\{z(e^z-1)\}$. Show that, if $0<|z|<2\pi$, the function can be expanded in the form

$$\frac{1}{z^2} - \frac{1}{2z} + a_0 + a_2z^2 + a_4z^4 + \ldots,$$

and find the values of a_0 and a_2.

10. The only singularities of a single-valued function $f(z)$ are poles of orders 1 and 2 at $z=-1$ and $z=2$, with residues at these poles 1 and 2 respectively. If $f(0)=7/4$, $f(1)=5/2$, determine the function and expand it in a Laurent series valid in $1<|z|<2$.

11. Classify the points $z=0$, $z=1$ and the point at infinity, in relation to the function

$$f(z) = \frac{z-2}{z^2}\sin\frac{1}{1-z},$$

and find the residues of $f(z)$ at $z=0$ and at $z=1$.

12. Show that, if b is real, the series

$$\tfrac{1}{2}\log(1+b^2)+i\arctan b + \frac{z-ib}{1+ib} - \tfrac{1}{2}\left(\frac{z-ib}{1+ib}\right)^2 + \tfrac{1}{3}\left(\frac{z-ib}{1+ib}\right)^3 - \ldots$$

is an analytic continuation of the function defined by the series

$$z - \tfrac{1}{2}z^2 + \tfrac{1}{3}z^3 - \dots .$$

13. The power series $z + \tfrac{1}{2}z^2 + \tfrac{1}{3}z^3 + \dots$ and

$$i\pi - (z-2) + \tfrac{1}{2}(z-2)^2 - \tfrac{1}{3}(z-2)^3 + \dots$$

have no common region of convergence : prove that they are nevertheless analytic continuations of the same function.

14. If $a > e$, use Rouché's theorem to prove that $e^z = az^n$ has n roots inside the circle $|z| = 1$.

15. *The Fundamental Theorem of Algebra.* By taking $f(z) = a_0 z^m$, $g(z) = a_1 z^{m-1} + a_2 z^{m-2} + \dots + a_m$, use Rouché's theorem to prove that the polynomial

$$F(z) = a_0 z^m + a_1 z^{m-1} + \dots + a_m$$

has exactly m zeros within the circle $|z| = R$ for sufficiently large R.

Deduce from Liouville's theorem that $F(z)$ has at least one zero.

16. Prove that $z^8 + 3z^3 + 7z + 5$ has exactly two zeros in the first quadrant.

17. If $|f(z)| > m$ on $|z| = a$, $f(z)$ is regular for $|z| \leqslant a$ and $|f(0)| < m$, prove that $f(z)$ has at least one zero in $|z| < a$. (See § **42**.)

Deduce that every algebraic equation has a root. (This is another proof of the Fundamental Theorem of Algebra.)

18. If a domain D of the z-plane is bounded by a simple closed contour C and $w = f(z)$ is regular in D and on C, prove that, if $f(z)$ takes no value more than once on C, then $f(z)$ takes no value more than once in D. (Use the theorem of § **40**.)

Prove that the above result holds for the function $w = z^2 + 2z + 3$, if D is the domain $|z| < 1$ and C is the unit-circle.

CHAPTER V

THE CALCULUS OF RESIDUES

§ 43. The Residue Theorem

We now turn our attention to the residue theorem, and to one of the first applications which Cauchy made of this theorem—the evaluation of definite integrals. It should be observed that a definite integral which can be evaluated by Cauchy's method of residues can also be evaluated by other means, though usually not so easily.*

We have already defined the residue of a function $f(z)$ at the pole $z = a$ to be the coefficient of $(z-a)^{-1}$ in the Laurent expansion of $f(z)$, which, if $z = a$ is a pole of order m, takes the form

$$\sum_0^\infty a_n(z-a)^n + \sum_1^m b_n(z-a)^{-n}.$$

We have also remarked that, when $z = a$ is a pole of order *one*, the residue b_1 can be calculated as $\lim_{z \to a}\{(z-a)f(z)\}$. The residue can also be defined as follows. If the point $z = a$ is the only singularity of $f(z)$ inside a closed contour C, and if $\dfrac{1}{2\pi i} \displaystyle\int_C f(z)dz$ has a value, that value is the residue of $f(z)$ at $z = a$.

The residue of $f(z)$ *at infinity* may also be defined. If $f(z)$ has an isolated singularity at infinity, or is regular

* For $\displaystyle\int_0^\infty e^{-x^2}\,dx$, sometimes stated to be an integral which cannot be evaluated by Cauchy's method, see Courant, *Differential and Integral Calculus*, II, p. 561. In this case Cauchy's method is the more difficult.

there, and if C is a large circle which encloses all the finite singularities of $f(z)$, then the residue at $z = \infty$ is defined to be

$$\frac{1}{2\pi i} \int_C f(z)dz$$

taken round C in the *negative* sense (negative with respect to the origin), provided that this integral has a definite value. If we apply the transformation $z = 1/\zeta$ to the integral, it becomes

$$\frac{1}{2\pi i} \int -f\left(\frac{1}{\zeta}\right) \frac{d\zeta}{\zeta^2}$$

taken positively round a small circle, centre the origin. It follows that if

$$\lim_{\zeta \to 0}\{-f\left(\frac{1}{\zeta}\right)/\zeta\} = \lim_{z \to \infty}\{-zf(z)\}$$

has a definite value, that value is the residue of $f(z)$ at infinity.

Note that a function may be regular at $z = \infty$ but yet have a residue there.

The function $f(z) = A/z$ has a residue A at $z = 0$ and a residue $- A$ at $z = \infty$, although $f(z)$ is regular at $z = \infty$.

Theorem 1.   Cauchy's Residue Theorem.

Let $f(z)$ be continuous within and on a closed contour C and regular, save for a finite number of poles, within C. Then

$$\int_C f(z)dz = 2\pi i \, \Sigma \mathcal{R},$$

where $\Sigma \mathcal{R}$ is the sum of the residues of $f(z)$ at its poles within C.

Let a_1, a_2, ..., a_n be the n poles within C.   Draw a set of circles γ_r of radius δ and centre a_r, which do not intersect and which all lie inside C.   Then $f(z)$ is certainly regular in the region between C and these small circles γ_r.

We can therefore deform C until it consists of the small circles γ_r and a polygon P which joins together the small circles as illustrated in fig. 13.

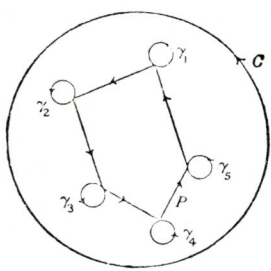

FIG. 13.

Then

$$\int_C f(z)dz = \int_P f(z)dz + \sum_{r=1}^{n} \int_{\gamma_r} f(z)dz = \sum_{r=1}^{n} \int_{\gamma_r} f(z)dz,$$

for the integral round the polygon P vanishes because $f(z)$ is regular within and on P.

If a_r is a pole of order m, then

$$f(z) = \phi(z) + \sum_{s=1}^{m} \frac{b_s}{(z-a_r)^s},$$

where $\phi(z)$ is regular within and on γ_r. Hence

$$\int_{\gamma_r} f(z)dz = \sum_{s=1}^{m} \int_{\gamma_r} \frac{b_s}{(z-a_r)^s}\, dz.$$

On writing $z = a_r + \delta e^{i\theta}$, θ varies from 0 to 2π as the point z makes a circuit of the circle γ_r, and so

$$\int_{\gamma_r} f(z)dz = \sum_{s=1}^{m} b_s \delta^{1-s} \int_0^{2\pi} e^{(1-s)i\theta} i\, d\theta = 2\pi i b_1.$$

Hence $$\int_C f(z)dz = \sum_{r=1}^{m} \int_{\gamma_r} f(z)dz = 2\pi i\, \Sigma \mathcal{R}.$$

which proves the theorem.

Theorem 2. *If* $\lim\limits_{z \to a}\{(z-a)f(z)\} = b$, *and if C is the arc,* $\theta_1 \leqslant \arg(z-a) \leqslant \theta_2$, *of the circle* $\mid z-a \mid = r$, *then*

$$\lim_{r \to 0} \int_C f(z)dz = ib(\theta_2 - \theta_1).$$

Given ϵ, we can find an $\eta(\epsilon)$ such that, if $\mid z-a \mid < \eta$, $\mid \delta \mid < \epsilon$, where $(z-a)f(z) = b + \delta$.

$$\int_C f(z)\,dz = \int_C \frac{b+\delta}{z-a}\,dz = i \int_{\theta_1}^{\theta_2} (b+\delta)d\theta.$$

Hence $$\left| \int_C f(z)dz - ib(\theta_2 - \theta_1) \right| < \epsilon(\theta_2 - \theta_1),$$

and so, on taking the limit as $r \to 0$, the theorem follows.

If $z = a$ is a *simple* pole of $f(z)$, b is the residue of $f(z)$ at $z = a$, so that if the contour is a small circle surrounding the pole, $\theta_2 - \theta_1 = 2\pi$ and we get

$$\int_C f(z)\,dz = 2\pi ib.$$

§ 44. Integration round the Unit Circle

We consider first the evaluation by contour integration of integrals of the type

$$\int_0^{2\pi} \phi(\cos\theta,\, \sin\theta)d\theta,$$

where $\phi(\cos\theta,\, \sin\theta)$ is a rational function of $\sin\theta$ and $\cos\theta$. If we write $z = e^{i\theta}$, then

$$\cos\theta = \frac{1}{2}\left(z + \frac{1}{z}\right),\ \ \sin\theta = \frac{1}{2i}\left(z - \frac{1}{z}\right),\ \ \frac{dz}{iz} = d\theta\,;$$

and so $$\int_0^{2\pi} \phi(\cos\theta,\, \sin\theta)d\theta = \int_C \psi(z)dz,$$

where $\psi(z)$ is a rational function of z and C is the unit circle $|z| = 1$. Hence

$$\int_C \psi(z)dz = 2\pi i \, \Sigma\mathcal{R}_C$$

where $\Sigma\mathcal{R}_C$ denotes the sum of the residues of $\psi(z)$ at its poles inside C.

Example.   Prove that, if $a > b > 0$,

$$J \equiv \int_0^{2\pi} \frac{\sin^2\theta \, d\theta}{a + b\cos\theta} = \frac{2\pi}{b^2}\{a - \sqrt{(a^2 - b^2)}\}.$$

Now on making the above change of variable, if C is the unit circle $|z| = 1$,

$$J = \frac{i}{2b}\int_C \frac{(z^2-1)^2 dz}{z^2(z^2 + 2az/b + 1)} = \frac{i}{2b}\int_C \frac{(z^2-1)^2 dz}{z^2(z-a)(z-\beta)} = \frac{i}{2b}\int_C F(z)dz,$$

where

$$a = \frac{-a + \sqrt{(a^2 - b^2)}}{b}, \quad \beta = \frac{-a - \sqrt{(a^2 - b^2)}}{b}$$

are the roots of the quadratic $z^2 + 2az/b + 1 = 0$. Since the product of the roots a, β is unity, we have $|a||\beta| = 1$ where $|\beta| > |a|$, and so $z = a$ is the only simple pole inside C. The origin is a pole of order two. We calculate the residues at (i) $z = a$, and (ii) $z = 0$.

(i) Residue $= \lim\limits_{z \to a}(z-a)F(z) = \lim\limits_{z \to a}\dfrac{(z^2-1)^2}{z^2(z-\beta)} = \dfrac{\left(a - \dfrac{1}{a}\right)^2}{a - \beta}$,

$$= \frac{(a-\beta)^2}{a-\beta} = a - \beta = \frac{2\sqrt{(a^2-b^2)}}{b}.$$

(ii) Residue is the coefficient of $1/z$ in $\dfrac{(z^2-1)^2}{z^2(z^2 + 2az/b + 1)}$, where z is small. Now

$$\frac{(z^2-1)^2}{z^2(z^2 + 2az/b + 1)} = \frac{1 - 2z^2 + \ldots}{z^2(1 + 2az/b + z^2)}$$

and coefficient of $1/z$ is plainly $-2a/b$.

Hence $\boldsymbol{J} = \dfrac{i}{2b} \cdot 2\pi i \, \Sigma \mathcal{R}_C = -\dfrac{\pi}{b}\left\{-\dfrac{2a}{b} + \dfrac{2\sqrt{(a^2-b^2)}}{b}\right\}$,

which proves the result.

§ 45. Evaluation of a Type of Infinite Integral

Let $Q(z)$ be a function of z satisfying the conditions :

(i) $Q(z)$ is meromorphic in the upper half-plane ;

(ii) $Q(z)$ has no poles on the real axis ;

(iii) $zQ(z) \to 0$ uniformly, as $|z| \to \infty$, for $0 \leqslant \arg z \leqslant \pi$;

(iv) $\displaystyle\int_0^\infty Q(x)dx$ and $\displaystyle\int_{-\infty}^0 Q(x)dx$ both converge ; then

$$\int_{-\infty}^\infty Q(x)dx = 2\pi i \, \Sigma \mathcal{R}^+$$

where $\Sigma \mathcal{R}^+$ denotes the sum of the residues of $Q(z)$ at its poles in the upper half-plane.

Choose as contour a semicircle, centre the origin and radius R, in the upper half-plane. Let the semicircle be denoted by Γ, and choose R large enough for the semicircle to include all the poles of $Q(z)$. Then, by the residue theorem,

$$\int_{-R}^R Q(x)dx + \int_\Gamma Q(z)dz = 2\pi i \, \Sigma \mathcal{R}^+.$$

From (iii), if R be large enough, $|zQ(z)| < \epsilon$ for all points on Γ, and so

$$\left|\int_\Gamma Q(z)dz\right| = \left|\int_0^\pi Q(Re^{i\theta})Re^{i\theta}id\theta\right| < \epsilon \int_0^\pi d\theta = \pi\epsilon.$$

Hence, as $R \to \infty$, the integral round Γ tends to zero. If (iv) is satisfied, it follows that

$$\int_{-\infty}^\infty Q(x)dx = 2\pi i \, \Sigma \mathcal{R}^+.$$

If $Q(z)$ be a rational function of z, it will be the ratio of two polynomials $N(z)/D(z)$, and condition (iv) is satisfied if the degree of $D(z)$ exceed that of $N(z)$ by at least *two*, for, when x is large, $Q(x)$ behaves like x^{-p}, where $p \geqslant 2$ and

$$\int_\lambda^\infty \frac{dx}{x^p} \text{ and } \int_{-\infty}^{-\mu} \frac{dx}{x^p} \text{ both exist.*}$$

Note that condition (iv) is required as well as (iii), for the condition $xQ(x) \to 0$ is not in itself sufficient to secure the convergence of $\int^\infty Q(x)dx$. This can be seen by taking $Q(x) = (x \log x)^{-1}$.

Example. Prove that, if $a > 0$,

$$\int_0^\infty \frac{dx}{x^4 + a^4} = \frac{\pi}{2\sqrt{2}a^3}.$$

If $z^4 + a^4 = 0$, we have $z^4 = a^4 e^{\pi i}$, and the simple poles of the integrand are at $ae^{\pi i/4}$, $ae^{3\pi i/4}$, $ae^{5\pi i/4}$, $ae^{7\pi i/4}$. Of these, only the first two are in the *upper* half-plane. The conditions of the theorem are plainly satisfied, and so

$$\int_{-\infty}^\infty \frac{dx}{x^4 + a^4} = 2\pi i \, \Sigma \{\text{Residues at } z = ae^{\pi i/4}, \, ae^{3\pi i/4}\}.$$

Let k denote any one of these, then $k^4 = -a^4$ and the residue at the simple pole $z = k$ is $\lim_{z \to k}\{(z-k)(z^4-k^4)^{-1}\}$. This may be evaluated by Cauchy's formula, as applied to the evaluation of limits of expressions of the indeterminate form † $0/0$, and so

$$\lim_{z \to k} \frac{z-k}{z^4-k^4} = \lim_{z \to k} \frac{1}{4z^3} = \frac{1}{4k^3} = -\frac{k}{4a^4}.$$

Hence

$$\int_{-\infty}^\infty \frac{dx}{x^4 + a^4} = -2\pi i \, \frac{1}{4a^4}(ae^{\pi i/4} + ae^{3\pi i/4}),$$

$$= -\frac{\pi i}{2a^3}(e^{\pi i/4} - e^{-\pi i/4}) = -\frac{\pi i}{2a^3} \, 2i \sin\frac{\pi}{4} = \frac{\pi}{\sqrt{2}a^3}.$$

* For the convergence of infinite integrals, see P.A., p. 193, or G.I., p. 77.

† P.A., p. 106.

Hence
$$\int_0^\infty \frac{dx}{x^4+a^4} = \frac{\pi}{2\sqrt{2a^3}}.$$

The theorem can be extended to the case in which $D(z) = 0$ has non-repeated real roots, so that $Q(z)$ has *simple* poles *on* the real axis. We now indent the contour by making small semicircles in the upper half-plane to cut out the simple poles on the real axis. Suppose that $D(z) = 0$ has only one root $z = a$, where a is real. The contour is then as shown in fig. 14. The small semicircle

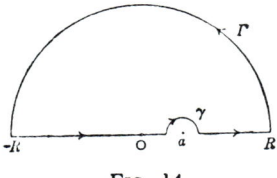

Fig. 14.

is denoted by γ, its centre is the point $x = a$ and its (small) radius is ρ. If Γ is large enough to enclose all the poles of $Q(z)$ in the upper half-plane, then the integral round Γ tends to zero as $R \to \infty$, as before. We therefore have, if the path of integration be as indicated by the arrows in fig 14,

$$\int_\Gamma + \int_{-R}^{a-\rho} + \int_\gamma + \int_{a+\rho}^R Q(z)dz = 2\pi i \, \Sigma \mathcal{R}^+.$$

As $R \to \infty$, $\int_{-R}^{a-\rho} + \int_{a+\rho}^R Q(x)dx = P \int_{-\infty}^\infty Q(x)dx,$

and it remains to consider $\int_\gamma Q(z)dz$. Now, on γ, $z = a + \rho e^{i\theta}$, and so

$$\int_\gamma Q(z)dz = \int_\pi^0 Q(a+\rho e^{i\theta})\rho e^{i\theta} i d\theta.$$

Since $Q(z)$ contains the factor $(z-a)^{-1}$ we may write $Q(z) = \phi(z)/(z-a)$ and $\phi(z)$ is regular at and near $z = a$. Hence

$$\int_\gamma Q(z)dz = \int_\pi^0 \phi(a+\rho e^{i\theta})id\theta = i\int_\pi^0 \{\phi(a)+O(\rho)\}d\theta,$$

since $\phi(a+\rho e^{i\theta})$ is regular at and near a and can be expanded by Taylor's theorem with remainder when $n = 1$. It follows that

$$\int_\gamma Q(z)dz \to -\pi i\phi(a) \text{ as } \rho \to 0.$$

Since $\phi(a)$ is plainly the residue of $Q(z) = \phi(z)/(z-a)$ at $z = a$, we can write the final result in the form

$$P\int_{-\infty}^\infty Q(x)dx = 2\pi i\, \Sigma\mathcal{R}^+ + \pi i\, \Sigma\mathcal{R}^0,$$

where $\Sigma\mathcal{R}^0$ denotes the sum of the residues of $Q(z)$ at its simple poles *on* the real axis, for clearly each pole on the real axis can be treated in the same way as $z = a$. The principal value of the integral is involved, because equal spaces ρ are taken on either side of the real poles, and, by definition,*

$$\lim_{\rho\to 0}\int_a^{a-\rho} + \int_{a+\rho}^\beta f(x)dx = P\int_a^\beta f(x)dx.$$

It should be noticed that if a pole be cut out by a small *semicircle*, the contribution to the value of the integral is half what it would be if a small *circle* surrounded the pole. (See § 43, Theorem 2.)

§ 46. Evaluation of Infinite Integrals by Jordan's Lemma

We now prove a very useful theorem which is usually known as *Jordan's lemma*.

* See P.A., p. 195, or G.I., p. 81.

If Γ be a semicircle, centre the origin and radius R, and $f(z)$ be subject to the conditions :

(i) $f(z)$ *is meromorphic in the upper half-plane,*
(ii) $f(z) \to 0$ *uniformly as* $|z| \to \infty$ *for* $0 \leqslant \arg z \leqslant \pi$,
(iii) m *is positive* ; *then*

$$\int_\Gamma e^{miz} f(z) dz \to 0 \text{ as } R \to \infty.$$

By (ii), if R is sufficiently large, we have, for all points on Γ, $|f(z)| < \epsilon$.   Now

$$|\exp miz| = |\exp\{miR(\cos\theta + i\sin\theta)\}| = \exp(-mR\sin\theta).$$
Hence,

$$\left| \int_\Gamma f(z)e^{miz}dz \right| = \left| \int_0^\pi f(z)e^{miz}Re^{i\theta}id\theta \right| < \epsilon \int_0^\pi e^{-mR\sin\theta} Rd\theta,$$
$$= 2R\epsilon \int_0^{\frac{1}{2}\pi} e^{-mR\sin\theta}d\theta.$$

Now it can be proved, by considering the sign of its derivative, or otherwise, that $\sin\theta/\theta$ steadily decreases from 1 to $2/\pi$ as θ increases from 0 to $\frac{1}{2}\pi$.   Hence, if $0 \leqslant \theta \leqslant \frac{1}{2}\pi$,

$$\frac{\sin\theta}{\theta} \geqslant \frac{2}{\pi}.$$

Hence

$$\left| \int_\Gamma f(z)e^{miz}dz \right| \leqslant 2R\epsilon \int_0^{\frac{1}{2}\pi} e^{-2mR\theta} \pi d\theta,$$
$$= \frac{\pi\epsilon}{m}(1 - e^{-mR}) < \frac{\pi\epsilon}{m},$$

from which the lemma follows.

By using this lemma, we can evaluate another type of integral.   The method may be set out as a theorem as follows.

Let $Q(z) = N(z)/D(z)$, where $N(z)$ and $D(z)$ are polynomials, and $D(z) = 0$ has no real roots, then if

(i) *the degree of $D(z)$ exceeds that of $N(z)$ by at least one,*
(ii) $m > 0$,

$$\int_{-\infty}^{\infty} Q(x)e^{mix}dx = 2\pi i \Sigma \mathscr{R}^{+},$$

where $\Sigma \mathscr{R}^{+}$ denotes the sum of the residues of $Q(z)e^{miz}$ at its poles in the upper half-plane.

If we write $f(z) = Q(z)e^{miz}$, we see that $f(z)$ satisfies the conditions of Jordan's lemma and so $\int_{\Gamma} f(z)dz \to 0$ as $R \to \infty$. On using the same contour as before, a large semicircle in the upper half-plane, by making $R \to \infty$ we get

$$\int_{-\infty}^{\infty} Q(x)e^{mix}dx = 2\pi i \Sigma \mathscr{R}^{+}.$$

On taking real and imaginary parts of this result we see that by this method we can evaluate integrals of the type

$$\int_{-\infty}^{\infty} f(x) \cos mx \, dx \, , \, \int_{-\infty}^{\infty} f(x) \sin mx \, dx.$$

By a well-known test for convergence of infinite integrals,* if $f(x)$ decreases and $\to 0$ as $x \to \infty$, since $\int_{a}^{x} \frac{\cos}{\sin} mx \, dx$ is bounded, the integrals in question converge.

Example. Prove that, if $a > 0$, $m > 0$,

$$\int_{0}^{\infty} \frac{\cos mx \, dx}{(a^2 + x^2)^2} = \frac{\pi}{4a^3} (1 + am)e^{-am}.$$

The only pole of the integrand considered, $e^{miz}(a^2 + z^2)^{-2}$, in the upper half-plane, is a double pole at $z = ai$. The

---

* The test is known as Dirichlet's test. See Titchmarsh, *Theory of Functions* (Oxford, 1932), p. 21.

conditions of the theorem are easily seen to be satisfied and so

$$\int_{-\infty}^{\infty} \frac{e^{mix}}{(a^2+x^2)^2}\, dx = 2\pi i\{\text{Residue of } \frac{e^{miz}}{(a^2+z^2)^2} \text{ at } z = ai\}.$$

Put $z = ai+t$ then, since t is small,

$$\frac{e^{miz}}{(a^2+z^2)^2} = \frac{e^{-ma}\, e^{mit}}{t^2(2ai+t)^2} = \frac{e^{-ma}}{-4a^2t^2}\, (1+mit+\dots) \left(1 - \frac{2t}{2ai}\, +\dots\right)$$

and the residue, which is the coefficient of t^{-1}, is easily seen to be

$$\frac{-ie^{-ma}(1+ma)}{4a^3}.$$

Hence
$$\int_{-\infty}^{\infty} \frac{e^{mix}dx}{(a^2+x^2)^2} = \frac{\pi e^{-ma}(1+ma)}{2a^3}.$$

On equating real parts, and taking one half of the result for the integral from 0 to ∞, we get

$$\int_{0}^{\infty} \frac{\cos mx \, dx}{(a^2+x^2)^2} = \frac{\pi}{4a^3}\, (1+am)e^{-am}.$$

If there are simple poles of the integrand on the real axis, we get a modification of this result, similar to that obtained for the theorem of § **45**. Thus, if $D(z) = 0$ *has non-repeated real roots*, we get

$$P\int_{-\infty}^{\infty} Q(x)e^{mix}dx = 2\pi i\Sigma\mathcal{R}^+ + \pi i\Sigma\mathcal{R}^0;$$

the proof following the same lines as before.

An example of this extension of the theorem is the proof that

$$\int_{0}^{\infty} \frac{\sin mx}{x}\, dx = \tfrac{1}{2}\pi, \text{ if } m>0.$$

On considering the integrand e^{mix}/z we see that it has a simple pole at $z = 0$ and none in the upper half-plane. The residue at $z = 0$ is easily seen to be unity, and so we get

$$P\int_{-\infty}^{\infty} \frac{e^{mix}}{x}\, dx = \pi i.$$

On equating real and imaginary parts we get

$$P \int_{-\infty}^{\infty} \frac{\cos mx}{x} \, dx = 0,$$

$$\int_{-\infty}^{\infty} \frac{\sin mx}{x} \, dx = \pi \, ;$$

the " P " is not necessary in the second integral, since $\sin mx / x \to m$ as $x \to 0$, whereas the integrand in the first integral becomes infinite at the origin. From the second result we get

$$\int_{0}^{\infty} \frac{\sin mx}{x} \, dx = \tfrac{1}{2}\pi.$$

§ 47. Integrals involving Many-Valued Functions

A type of integral of the form $\int_{0}^{\infty} x^{a-1} Q(x) dx$, where a is not an integer, can also be evaluated by contour integration, but since z^{a-1} is a many-valued function, it becomes necessary to use the cut plane. One method of dealing with integrals of this type is to use as contour a large circle Γ, centre the origin, and radius R; but we must cut the plane along the real axis from 0 to ∞ and also enclose the branch-point $z = 0$ in a small circle γ of radius ρ. The contour is illustrated in fig. 15.

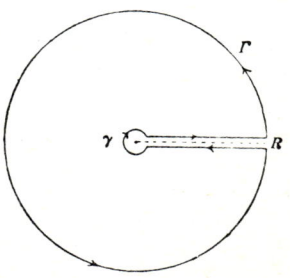

Fig. 15.

Let $Q(x)$ be a rational function of x with no poles on the real axis. If we write $f(z) = z^{a-1}Q(z)$ and suppose that $zf(z) \to 0$ uniformly both as $|z| \to 0$ and as $|z| \to \infty$, then we get the integral round Γ tending to zero as $R \to \infty$ and the integral round γ tending to zero as $\rho \to 0$; for, on Γ, if R is large enough, $|zf(z)| < \epsilon$ and so $|f(z)| < \epsilon/R$,

$$\left| \int_\Gamma f(z)dz \right| < \frac{\epsilon}{R} 2\pi R = 2\pi\epsilon.$$

Similarly on γ, $|zf(z)| < \epsilon$ if ρ is small enough, and so $|f(z)| < \epsilon/\rho$ and

$$\left| \int_\gamma f(z)dz \right| < \frac{\epsilon}{\rho} 2\pi\rho = 2\pi\epsilon.$$

Hence on making $\rho \to 0$ and $R \to \infty$ we get

$$\int_0^\infty x^{a-1}Q(x)dx + \int_\infty^0 x^{a-1}e^{2\pi i(a-1)}Q(x)dx = 2\pi i \Sigma \mathcal{R},$$

where $\Sigma \mathcal{R}$ is sum of residues of $f(z)$ inside the contour. We observe that the values of x^{a-1} at points on the upper and lower edges of the cut are not the same, for, if $z = re^{i\theta}$, we have $z^{a-1} = r^{a-1}e^{i\theta(a-1)}$ and the values of z at points on the upper edge correspond to $|z| = r$, $\theta = 0$, and at points on the lower edge they correspond to $|z| = r$, $\theta = 2\pi$.

Since $e^{2\pi i(a-1)} = e^{2\pi i a}$, we get

$$\int_0^\infty x^{a-1}Q(x)dx = \frac{2\pi i \Sigma \mathcal{R}}{1 - e^{2\pi i a}}.$$

We also observe that, when calculating the residues at the poles, z^{a-1} must be given its correct value $r^{a-1}e^{i\theta(a-1)}$ at each pole.

Example.   Prove that

$$\int_0^\infty \frac{x^{a-1}dx}{1+x} = \frac{\pi}{\sin a\pi}, \text{ if } 0 < a < 1.$$

Here we observe that, when $f(z) = z^{a-1}(1+z)^{-1}$, $zf(z) \to 0$ as $|z| \to \infty$, if $a < 1$, and $zf(z) \to 0$ as $|z| \to 0$, if $a > 0$. Hence, if $0 < a < 1$, the integral round Γ tends to zero as $R \to \infty$ and the integral round γ tends to zero as $\rho \to 0$.

Thus

$$\int_0^\infty \frac{x^{a-1}}{1+x}\, dx = \frac{2\pi i}{1 - e^{2\pi i a}} \{\text{Residue of } z^{a-1}(1+z)^{-1} \text{ at } z = -1\}.$$

At $z = -1$ we have $r = 1$, $\theta = \pi$, and so, for the residue,

$$\lim_{z \to -1} \left\{ (1+z)\, \frac{z^{a-1}}{z+1} \right\} = (-1)^{a-1} = e^{(a-1)\pi i} = -e^{a\pi i}.$$

Hence

$$\int_0^\infty \frac{x^{a-1}}{1+x}\, dx = -2\pi i \left\{ \frac{e^{a\pi i}}{1 - e^{2\pi a i}} \right\} = -2\pi i \frac{1}{e^{-a\pi i} - e^{a\pi i}}$$

$$= \frac{\pi}{\sin a\pi}.$$

This integral can also be evaluated by integrating $z^{a-1}/(1-z)$, using as contour a large semicircle in the upper half plane and the real axis indented by semicircles at $z = 0$ and at $z = 1$. In this case the cut plane is unnecessary. The evaluation of the integral by this second method is left as an exercise for the reader.* By this second method one obtains the further result that

$$P \int_0^\infty \frac{x^{a-1}\, dx}{1-x} = \pi \cot a\pi.$$

§ 48. Use of Contour Integration for deducing Integrals from Known Integrals

The contours used so far have been either circles or semicircles, and although a large semicircle in the upper half plane is generally used for integrals of the type discussed in § 45, there is no special merit in a semicircle. The rectangle with vertices $\pm R$, $\pm R + iR$ could also be

* See Copson, *Functions of a Complex Variable*, p. 140.

used in these cases. We now give two examples of deducing the values of some useful integrals by integrating a given function round a prescribed contour. We use, in the first case, a rectangle and in the second a quadrant of a circle.

Example 1. *Prove that* $\int_0^\infty e^{-x^2} \cos 2ax\, dx = \frac{1}{2}\sqrt{\pi} e^{-a^2}$ *by integrating* e^{-z^2} *round the rectangle whose vertices are* 0, *R*, *R*+*ia*, *ia*.

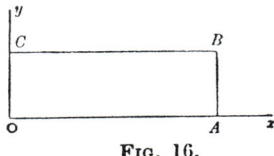

Fig. 16.

Let *A* be (*R*, 0) and *C* be (0, *a*) in the Argand diagram. On *OA*, $z = x$; on *AB*, $z = R + iy$; on *BC*, $z = x + ia$; and on *OC*, $z = iy$. Now e^{-z^2} has no poles within or on this contour and so, by Cauchy's theorem,

$$\int_0^R e^{-x^2}dx + \int_0^a e^{-(R+iy)^2}idy + \int_R^0 e^{-(x+ia)^2}dx + \int_a^0 e^{y^2}idy = 0.$$

Hence

$$\int_0^R e^{-x^2}dx - e^{a^2}\int_0^R e^{-x^2}(\cos 2ax - i\sin 2ax)dx +$$
$$i\int_0^a e^{-R^2}e^{-2iRy+y^2}dy - i\int_0^a e^{y^2}dy = 0. \qquad . \quad (1)$$

Now $\left| i\int_0^a e^{-R^2}e^{-2iRy+y^2}dy \right| < e^{-R^2}.\ e^{a^2}.\ a$

and so this integral $\to 0$ as $R \to \infty$. On using the result that

$$\int_0^\infty e^{-x^2}dx = \frac{1}{2}\sqrt{\pi},$$

we find, on making $R \to \infty$ and equating real parts in (1),

$$\int_0^\infty e^{-x^2}\cos 2ax\, dx = \frac{1}{2}\sqrt{\pi}\, e^{-a^2}.$$

Example 2. By integrating $e^{iz}z^{a-1}$ round a quadrant of a circle of radius R, prove that, if $0 < a < 1$,

$$\int_0^\infty x^{a-1} \frac{\cos}{\sin} x \, dx = \Gamma(a) \frac{\cos}{\sin} \frac{\pi a}{2}.$$

The contour required is drawn in fig 17.

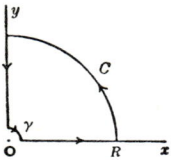

FIG. 17.

Since the origin is a branch-point for the function z^{a-1}, we enclose it in a quadrant of a small circle γ of radius ρ. We integrate round the contour in the sense indicated by the arrows. On γ, $z = \rho e^{i\theta}$ and we get

$$\left| \int_\gamma e^{iz}z^{a-1}dz \right| = \left| \int_{\frac{1}{2}\pi}^0 e - \rho\sin\theta \rho^{a-1} e^{(a-1)i\theta} \rho e^{i\theta}id\theta \right|,$$

$$\leqslant \rho^a \left| \int_{\frac{1}{2}\pi}^0 d\theta \right| = \tfrac{1}{2}\pi\rho^a,$$

since $|e^{-\rho\sin\theta}| \leqslant 1$ when ρ is small. It follows that

$$\int_\gamma e^{iz}z^{a-1}dz \to 0 \text{ as } \rho \to 0, \text{ if } a > 0.$$

If $a < 1$, $|z^{a-1}| < \epsilon$ when R is large enough, and so by the same argument as was used in proving Jordan's lemma (§ 46) we have

$$\int_C e^{iz}z^{a-1}dz \to 0 \text{ as } R \to \infty, \text{ if } a < 1.$$

Hence, if $0 < a < 1$, we get, on making $\rho \to 0$ and $R \to \infty$,

$$\int_0^\infty e^{ix}x^{a-1}dx + \int_\infty^0 e^{-y}y^{a-1}e^{(a-1)i\pi/2}idy = 0,$$

since there are no poles inside the contour.   Hence

$$\int_0^\infty x^{a-1}(\cos x + i \sin x)dx = \int_0^\infty e^{-y}y^{a-1}\left(\cos \frac{\pi a}{2} + i \sin \frac{\pi a}{2}\right)dy.$$

Since $\displaystyle\int_0^\infty e^{-y}y^{a-1}dy = \Gamma(a),$* on equating real and imaginary

parts, the required results follow.

§ 49. Expansion of a Meromorphic Function

Let $f(z)$ be a function whose only singularities, except at infinity, are simple poles at the points $z = a_1$, $z = a_2$, $z = a_3$, ... ; and suppose that

$$0 < |a_1| < |a_2| < |a_3| < \dots.$$

Suppose also that we know the residues at these poles : let them be b_1, b_2, b_3, ...   Consider a sequence of closed contours, either circles or squares, C_1, C_2, C_3, ..., such that C_n encloses a_1, a_2, ..., a_n but no other poles.   The contours C_n must be such that (i) the minimum distance R_n of C_n from the origin tends to infinity with n, (ii) the length L_n of the contour C_n is $O(R_n)$, (iii) on C_n we must have $f(z) = o(R_n)$.   Condition (iii) would be satisfied if $f(z)$ were bounded on the whole system of contours C_n.

When these conditions are satisfied we can prove that, *for all values of z except the poles themselves,*

$$f(z) = f(0) + \sum_{n=1}^\infty b_n\left(\frac{1}{z-a_n} + \frac{1}{a_n}\right).$$

To prove this, consider the integral

$$J \equiv \frac{1}{2\pi i}\int_{C_n} \frac{f(\zeta)}{\zeta(\zeta-z)}\,d\zeta,$$

where z is a point within C_n.   The integrand has poles at the points a_m with residues $b_m/\{a_m(a_m - z)\}$;   at $\zeta = z$

* G.I., p. 84.

with residue $f(z)/z$; and at $\zeta = 0$ with residue $-f(0)/z$. In particular cases the last two residues may be zero. Hence

$$J = \sum_{m=1}^{n} \frac{b_m}{a_m(a_m - z)} - \frac{f(0)}{z} + \frac{f(z)}{z}.$$

If now we can prove that $J \to 0$ as $n \to \infty$, the theorem is proved. Here we require the conditions laid down above on the contours C_n. On making use of these, we see that

$$|J| \leqslant \frac{L_n}{2\pi R_n(R_n - |z|)} \max_{C_n} |f(\zeta)| \to 0 \text{ as } n \to \infty.$$

The series is uniformly convergent in any finite region which does not contain any of the poles.

As an example of this theorem we prove that

$$\operatorname{cosec} z - \frac{1}{z} = 2z \sum_{n=1}^{\infty} \frac{(-)^{n-1}}{n^2 \pi^2 - z^2}.$$

Consider the function $f(z) = \operatorname{cosec} z - \dfrac{1}{z}$ $(z \neq 0), f(0) = 0$. Now $\sin z$ has simple zeros at the points $z = n\pi, (n = \ldots -2, -1, 1, 2, \ldots)$ and so $f(z) = \dfrac{z - \sin z}{z \sin z}$ will have simple poles at these points. The residue at $z = n\pi$ becomes, on writing $z - n\pi = \zeta$,

$$\lim_{\zeta \to 0} \frac{\zeta\{\zeta + n\pi - \sin(\zeta + n\pi)\}}{(\zeta + n\pi)\sin(\zeta + n\pi)} = \lim_{\zeta \to 0} \frac{\zeta + n\pi - \sin(\zeta + n\pi) + \zeta\{1 - \cos(\zeta + n\pi)\}}{(\zeta + n\pi)\cos(\zeta + n\pi) + \sin(\zeta + n\pi)}$$

$$= \frac{n\pi}{n\pi \cos n\pi} = (-1)^n.$$

There is no singularity at $z = 0$ since

$$\frac{z - \sin z}{z \sin z} = \frac{O\{|z|^3\}}{z^2 + O\{|z|^4\}} = O(|z|).$$

Let C_n be the square with corners at the points $(n + \frac{1}{2})(\pm 1 \pm i)\pi$. The function $1/z$ is certainly bounded on these squares. To prove that $\operatorname{cosec} z$ is also bounded, consider separately the regions (a) $y > \frac{1}{2}\pi$, (b) $y < -\frac{1}{2}\pi$, (c) $-\frac{1}{2}\pi \leqslant y \leqslant \frac{1}{2}\pi$. In (a) we have $y > \frac{1}{2}\pi$ and

$$| \operatorname{cosec} z | = \left| \frac{2}{e^{iz}-e^{-iz}} \right| \leqslant \frac{2}{e^{\frac{1}{2}\pi}-e^{-\frac{1}{2}\pi}},$$

for $| e^{iz}-e^{-iz} | \geqslant |\{ | e^{iz} | - | e^{-iz} | \}| = |\{ | e^{-y} | - | e^{y} | \}| ;$

and a similar argument applies to (b), writing $y = -t$ so that $t > \frac{1}{2}\pi$. For (c), let AB be the line joining the points $\frac{1}{2}\pi \pm \frac{1}{2}\pi i$.

Since $|\sin z| = (\cosh^2 y - \cos^2 x)^{\frac{1}{2}}$; on AB we have, since $x = \frac{1}{2}\pi$, $|\sin z| = \cosh y \geqslant 1$, so that $| \operatorname{cosec} z | \leqslant 1$.

Since $\operatorname{cosec} z$ has period π, it is bounded on all the lines joining $(n+\frac{1}{2}-\frac{1}{2}i)\pi$ and $(n+\frac{1}{2}+\frac{1}{2}i)\pi$. Hence $\operatorname{cosec} z$ is bounded on all the squares C_n. The previous theorem therefore gives

$$\operatorname{cosec} z - \frac{1}{z} = \sum_{n=-\infty}^{\infty}{}' (-1)^n \left(\frac{1}{z-n\pi} + \frac{1}{n\pi} \right),$$

the accent indicating that the term $n = 0$ is omitted from the summation. Since the series with $n > 0$ and with $n < 0$ converge separately, we may add together the terms corresponding to $\pm n$ and write the expansion

$$\operatorname{cosec} z - \frac{1}{z} = 2z \sum_{n=1}^{\infty} \frac{(-1)^{n-1}}{n^2\pi^2-z^2}.$$

§ 50. Summation of Series by the Calculus of Residues

The method of contour integration can be used with advantage for summing series of the type $\Sigma f(n)$, if $f(z)$ be a meromorphic function of a fairly simple kind.

Let C be a closed contour including the points m, $m+1, ..., n$, and suppose that $f(z)$ has simple poles at the points $a_1, a_2, ..., a_k$, with residues $b_1, b_2, ..., b_k$. Consider the integral

$$\int_C \pi \cot \pi z\, f(z)\, dz.$$

The function $\pi \cot \pi z$ has simple poles inside C at the points $z = m,\ m+1,\ ...,\ n$, with residue unity at each pole. The residues at these poles of $\pi \cot \pi z\, f(z)$ are accordingly $f(m),\ f(m+1),\ ...,\ f(n)$. Hence, by the residue theorem,

$$\int_C f(z)\, \pi \cot \pi z\ dz = 2\pi i\{f(m)+f(m+1)+...+f(n)$$
$$+b_1\pi \cot \pi a_1+...+b_k\pi \cot \pi a_k\}.$$

If conditions are satisfied which ensure that the contour integral tends to zero as $n \to \infty$, we can find the sum of the series $\Sigma f(n)$. Suppose that $f(z)$ is a rational function, none of whose zeros or poles are integers, such that $zf(z) \to 0$ as $|z| \to \infty$. Let C be the square with corners $(n+\tfrac{1}{2})(\pm 1 \pm i)$. We have seen that $\cot \pi z$ is bounded on this square and so $\left| \int_C zf(z)\pi \cot \pi z\, \dfrac{dz}{z} \right| \leqslant \dfrac{\pi M L\epsilon}{R}$ for n large enough, where M is the upper bound of $|\cot \pi z|$ on C, L is the length of C and R is the least distance of the origin from the contour. Since $L = 8R$, the integral tends to zero as $n \to \infty$, and so

$$\lim_{n \to \infty} \sum_{m=-n}^{n} f(m) = -\pi\{b_1 \cot \pi a_1+...+b_k \cot \pi a_k\}.$$

If we use $\pi \operatorname{cosec} \pi z$ instead of $\pi \cot \pi z$, we can obtain similarly the sums of series of the type $\Sigma(-1)^m f(m)$.

Example. Find the sums of the series $\displaystyle\sum_{n=1}^{\infty} \frac{1}{n^2+a^2},\ \sum_{n=0}^{\infty} \frac{(-1)^n}{n^2+a^2}.$
For the first, $f(z) = \dfrac{1}{z^2+a^2}$ and so $zf(z) \to 0$ as $|z| \to \infty$. The two poles of $f(z)$ are at $z = \pm ai$ and the residues at these poles are $\pm 1/2ai$. Hence

$$\sum_{m=-\infty}^{\infty} \frac{1}{m^2+a^2} = -\pi\left\{\frac{1}{2ai}\cot \pi ai - \frac{1}{2ai}\cot(-\pi ai)\right\} = \frac{\pi}{a}\coth \pi a,$$

or
$$\frac{1}{a^2} + 2\sum_{m=1}^{\infty} \frac{1}{m^2+a^2} = \frac{\pi}{a}\coth \pi a.$$

Similarly we get, by using $\pi \operatorname{cosec} \pi z$ instead of $\pi \cot \pi z$,

$$\sum_{m=0}^{\infty} \frac{(-1)^m}{m^2+a^2} = \frac{1}{2a^2} + \frac{\pi}{2a} \operatorname{cosech} \pi a.$$

In simple cases we can deal similarly with functions $f(z)$ which have poles which are not simple. As an example, consider the series

$$\sum_{-\infty}^{\infty} \frac{1}{(a+n)^2}.$$

Here $f(z) = (a+z)^{-2}$ has a double pole at $z = -a$. By Taylor's theorem

$$\cot \pi z = \cot(-\pi a) + (\pi z + \pi a)\{-\operatorname{cosec}^2(-\pi a)\} + \cdots,$$

and so the residue of $\cot \pi z/(z+a)^2$ at $z = -a$ is $-\pi \operatorname{cosec}^2 \pi a$.

Hence
$$\sum_{-\infty}^{\infty} \frac{1}{(a+n)^2} = \pi^2 \operatorname{cosec}^2 \pi a.$$

EXAMPLES V

Use the method of contour integration to prove the following results 1 to 10 :—

1. $\int_0^{\pi} \dfrac{a\, d\theta}{a^2+\sin^2\theta} = \dfrac{\pi}{\sqrt{(1+a^2)}}$, $(a>0)$.

2. $\int_0^{2\pi} \dfrac{\cos^2 3\theta\, d\theta}{1-2p \cos 2\theta + p^2} = \pi\,\dfrac{1-p+p^2}{1-p}$, $(0<p<1)$.

3. $\int_0^{2\pi} \dfrac{(1+2\cos\theta)^n \cos n\theta\, d\theta}{3+2\cos\theta} = \dfrac{2\pi}{\sqrt{5}}\,(3-\sqrt{5})^n$, $(n$ a positive integer$)$.

4. $\int_{-\infty}^{\infty} \dfrac{dx}{(x^2+b^2)(x^2+c^2)^2} = \dfrac{\pi(b+2c)}{2bc^3(b+c)^2}$, $(b>0,\ c>0)$.

5. $\int_0^{\infty} \dfrac{x^6 dx}{(a^4+x^4)^2} = \dfrac{3\sqrt{2}\,\pi}{16a}$, $(a>0)$.

6. $\int_{-\infty}^{\infty} \dfrac{\cos x\, dx}{(x^2+a^2)(x^2+b^2)} = \dfrac{\pi}{a^2-b^2}\left(\dfrac{e^{-b}}{b} - \dfrac{e^{-a}}{a}\right)$, $(a>b>0)$.

7. $\int_0^\infty \dfrac{\cos ax\,dx}{1+x^2+x^4} = \dfrac{\pi}{\sqrt{3}}\sin\tfrac{1}{2}\left(a+\dfrac{\pi}{3}\right)\exp\left(-\tfrac{1}{2}a\sqrt{3}\right),\ (a>0).$

8. $\int_0^\infty \dfrac{\sin^2 mx\,dx}{x^2(a^2+x^2)} = \dfrac{\pi}{4a^3}\left(e^{-2ma}-1+2ma\right),\ (m>0,\ a>0).$

9. $\int_0^\infty \dfrac{x^{a-1}dx}{1+x+x^2} = \dfrac{2\pi}{\sqrt{3}}\cos\left(\dfrac{2\pi a+\pi}{6}\right)\operatorname{cosec}\pi a,\ (0<a<2).$

10. $\int_0^\infty \dfrac{x^a dx}{(1+x^2)^2} = \dfrac{\pi(1-a)}{4\cos\tfrac{1}{2}\pi a},\ (-1<a<3).$

11. Evaluate $\int \dfrac{dz}{1+z^4}$ taken round the ellipse whose equation is $x^2-xy+y^2+x+y=0$. Evaluate similarly $\int \dfrac{dz}{1+z^3}$ round the ellipse $2x^2+y^2=4x$.

12. Show that the function $f(z) = z/(a-e^{-iz})$ has simple poles at the points $z=i\log a+2\pi n$, $(n=0,\ \pm1,\ \pm2,\ \ldots)$; and by integrating $f(z)$ round a rectangle with corners at $\pm\pi,\ \pm\pi+in$ prove that, if $a>1$,

$$\int_0^\pi \frac{x\sin x\,dx}{1+a^2-2a\cos x} = \frac{\pi}{a}\log\frac{1+a}{a}.$$

13. By taking as contour a square whose corners are $\pm N,\ \pm N+2Ni$, where N is an integer, and making $N\to\infty$, prove that

$$\int_0^\infty \frac{dx}{(1+x^2)\cosh(\tfrac{1}{2}\pi x)} = \log 2.$$

14. By integrating $e^{-kz}z^{n-1}$ round a sector, of radius R, bounded by the lines $\arg z=0$, $\arg z=a<\tfrac{1}{2}\pi$ (indented at 0), prove that, if $k>0,\ n>0$,

$$\int_0^\infty x^{n-1}e^{-kx\cos a}\,{\cos\atop\sin}\,(kx\sin a)dx = k^{-n}\Gamma(n)\,{\cos\atop\sin}\,na.$$

15. Prove that $P\int_0^\infty \dfrac{x^4 dx}{x^6-1} = \dfrac{\pi\sqrt{3}}{6}$.

16. Prove that $\int_0^\infty \dfrac{x-\sin x}{x^3(a^2+x^2)}\,dx = \dfrac{\pi}{2a^4}(\tfrac{1}{2}a^2-a+1-e^{-a}),\ (a>0).$

17. By integrating $e^{as}/(e^{-2is}-1)$ round a suitable contour, prove that

$$\int_0^\infty \frac{\sin ay\, dy}{e^{2y}-1} = \tfrac{1}{4}\pi \coth \tfrac{1}{2}\pi a - \frac{1}{2a}.$$

18. Prove that $\sec z = 4\pi \overset{\infty}{\underset{0}{\Sigma}}(-1)^n(2n+1)/\{(2n+1)^2\pi^2-4z^2\}$.

19. Prove that, if $-\pi < a < \pi$,

$$\frac{\sin az}{\sin \pi z} = \frac{2}{\pi} \overset{\infty}{\underset{n=1}{\Sigma}} (-1)^n \frac{n \sin na}{z^2-n^2},$$

$$\frac{\cos az}{\sin \pi z} = \frac{1}{\pi z} + \frac{2z}{\pi} \overset{\infty}{\underset{n=1}{\Sigma}} (-1)^n \frac{\cos na}{z^2-n^2}.$$

20. Prove that

$$\text{(i)} \quad \overset{\infty}{\underset{n=-\infty}{\Sigma}} \frac{1}{n^4+a^4} = \frac{\pi}{a^3\sqrt{2}} \frac{\sinh(\pi a\sqrt{2})+\sin(\pi a\sqrt{2})}{\cosh(\pi a\sqrt{2})-\cos(\pi a\sqrt{2})},$$

$$\text{and find} \quad \overset{\infty}{\underset{n=-\infty}{\Sigma}} \frac{(-1)^n}{n^4+a^4}.$$

(ii) Prove that

$$\overset{\infty}{\underset{n=1}{\Sigma}} \frac{1}{n^4-a^4} = \frac{1}{2a^4} - \frac{\pi}{4a^3}(\coth \pi a + \cot \pi a).$$

21. By integrating $\dfrac{\pi \sin az}{z^3 \sin \pi z}$ round a suitable contour, prove that

$$\frac{1}{1^3} - \frac{1}{3^3} + \frac{1}{5^3} - \frac{1}{7^3} + \ldots = \frac{\pi^3}{32}.$$

22. (i) Prove that $\int z^2 \log \dfrac{z+1}{z-1} dz$, taken round the circle $|z|=2$, has the value $4\pi i/3$.

(ii) Prove that $\int_0^\infty \dfrac{\log x\, dx}{(1+x^2)^2} = -\dfrac{\pi}{4}$, using as contour a large semicircle in the upper half-plane indented at the origin.

MISCELLANEOUS EXAMPLES

1. If $w = u + iv$ is a regular function of z, show that $\left(\dfrac{\partial^2}{\partial x^2} + \dfrac{\partial^2}{\partial y^2}\right) w = 0$ is equivalent to

$$\frac{\partial^2 w}{\partial z \partial \bar{z}} = 0.$$

Hence show that

$$u\left\{\frac{1}{2}(z+\bar{z}), \ \frac{1}{2i}(z-\bar{z})\right\} = \phi(z) + \psi(\bar{z}),$$

$$iv\left\{\frac{1}{2}(z+\bar{z}), \ \frac{1}{2i}(z-\bar{z})\right\} = \phi(z) - \psi(\bar{z}) + C,$$

where C is an arbitrary constant.*

Use the above relations to find the regular function $f(z)$ for which u is (i) $\log (x^2 + y^2)$, (ii) $x^2 - y^2 + 4xy$ and for which v is

$$e^{2x} (y \cos 2y + x \sin 2y).$$

2. If $x = r \cos \theta$, $y = r \sin \theta$ change the independent variables in $\left(\dfrac{\partial^2}{\partial x^2} + \dfrac{\partial^2}{\partial y^2}\right) \phi = 0$, to r, θ.

If $u = (x^2 + y^2)^2/(x^2 - y^2)$ find the function $\phi(u)$ which satisfies $\nabla^2 \phi = 0$. Find also the regular function $f(z)$ of which $\phi(u)$ is the real part.

3. Show that, if $\gamma \neq 0$, there are two points unaltered by the transformation

$$w = \frac{az + \beta}{\gamma z + \delta}$$

unless $(\delta - a)^2 + 4\beta\gamma = 0$, in which case there is only one such point. Show that, if $z = 1$ is this point,

$$\frac{1}{w-1} = \frac{1}{z-1} + \kappa$$

where κ is a constant.

* This result was communicated to me by Prof. A. Oppenheim.

Show that $z = 1$ is the only fixed point of the transformation

$$w = \frac{z - 1 + iz \tan \lambda}{z - 1 + i \tan \lambda},$$

$(0 < \lambda < \frac{1}{4}\pi)$, and that this transformation maps the inside of $|z| = 1$ on the inside of $|w| = 1$.

Sketch the curve in the w-plane which corresponds to the straight line joining $z = 1$ and $z = i$.

4. Prove that, if $k > 0$, $w = \tan (\pi z/4k)$ maps lines parallel to the axes in the z-plane on systems of coaxal circles in the w-plane.

Find what corresponds to the infinite strip $z = \pm k$ and indicate in a figure the region of the w-plane corresponding to the square $0 \leqslant x \leqslant k$, $k \leqslant y \leqslant 2k$.

5. Show that, if $a > 0$, the relation $w = ai \cot \frac{1}{2}z$ maps the semi-infinite strip $0 \leqslant x \leqslant 2\pi$, $y \geqslant 0$ on a half-plane cut from $u = a$ to $u = \infty$.

Two circles, with real limiting points $(\pm a, 0)$, are drawn in the cut w-plane with centres $(2ka, 0)$, $(ka, 0)$, where $k > 1$. Show that the space between these circles is mapped on the interior of a rectangle in the z-plane whose area is

$$\pi \log \frac{(2k-1)(k+1)}{(2k+1)(k-1)}.$$

6. Express the transformation

$$w = \frac{4(z^2 + 1)}{z^2 - 6z + 1}$$

in the form $\dfrac{w - a}{w - \beta} = k\left(\dfrac{z - a}{z - b}\right)^2$ and hence show that the inside of $|z| = 1$ is mapped on the whole w-plane cut along a segment of the real axis.

Illustrate by a diagram what corresponds in the w-plane to that part of the circle $|z - i| = \sqrt{2}$ which lies in the fourth quadrant.

7. If a, b, c, d are real constants, some of which may be zero, and

$$w = \frac{az^2 + bz + c}{z + d},$$

show that there are two values of w, each of which corresponds to a pair of equal values of z; and that these values of w can only be equal if the transformation is bilinear.

Discuss the transformation

$$w = \frac{z^2 - 2z}{1 - 2z}$$

in this way, and find the boundaries in the z-plane which correspond to $|w| = 1$.

Show in a diagram the regions of the z-plane corresponding to $|w| \leqslant 1$.

8. If $f(z)$ is regular within and on the circle $|z| = R$ and if $|f(z)| < M$ on the circle, prove that, if $|z| = r < R$,

$$|f(z)| > |a_0| - \frac{Mr}{R-r},$$

where a_0 is the constant term in $f(z) = \sum_{s=0}^{\infty} a_s z^s$.

If $R = 1$, $a_0 = 1$ and $|f(z)| < k$ on $|z| = 1$, prove that $f(z)$ does not vanish within the circle $|z| = 1/(1+k)$.

9. Show that, if $0 < a < \tfrac{1}{4}\pi$ and

$$f(z) = \frac{4az \cot a}{1 + 2z \cot a - z^2},$$

then $w = f(z)$ gives a conformal transformation when z lies in any finite region excluding the points $z = \pm i$, $z = \cot \tfrac{1}{2}a$, $z = -\tan \tfrac{1}{2}a$.

Show that the boundary of the semicircle $|z| = 1$, $\mathbf{R}z > 0$ corresponds to an arc of a circle in the w-plane subtending an angle $4a$ at the centre.

Find the two points in the z-plane corresponding to the centre of this circle.

10. Show that, if

$$w = \left(\frac{z+1}{z}\right)^2,$$

two finite points of the z-plane are mapped on every finite point of the w-plane, except the origin and $w = 1$, and explain why the mapping ceases to be conformal at the point $z = -1$.

Show, in a diagram, the two domains of the z-plane which are mapped on the semi-circular domain $|w| < 1$, $\mathbf{I}w > 0$.

11. If $f(z) = \sum\limits_{0}^{\infty} a_n z^n$ and $\phi(z) = \sum\limits_{0}^{\infty} b_n z^n$ are both regular within a domain D which includes the circle $|z| = 1$, prove that

$$\sum_{0}^{\infty} a_n b_n = \frac{1}{2\pi} \int_0^{2\pi} f(e^{i\theta}) \phi(e^{-i\theta}) d\theta.$$

Hence, or otherwise, prove that

$$\frac{1}{(2\,!)^2} - \frac{1}{(3\,!)^2} + \frac{1}{(4\,!)^2} - \ldots = \frac{1}{2\pi} \int_0^{2\pi} \cos(2 \sin \theta) d\theta.$$

12. A function $\phi(z)$ is regular over the whole z-plane, except at $z = 0$ and at $z = \infty$, and, for all values of z, $\phi(z) = z\phi(\beta z)$, where $|\beta| < 1$. Find the Laurent expansion of $\phi(z)$, given that the constant term in the expansion is unity.
   Show that $\phi(-\beta) = 0$.

13. $ABCD$ is a square whose vertices are $0, i, -1+i, -1$; Γ_1 is the line BC and Γ_2 consists of the other three sides of the square. If $f(z) = z^5 + z^2 + 1$ prove that (i) $f(z)$ does not assume any real non-negative value on Γ_1 (ii) $\mathbf{R} f(z) > 0$ at all points of Γ_2 except B.

Hence, or otherwise, evaluate

$$\int_{\Gamma_1} \frac{f'(z)}{f(z)} dz + \int_{\Gamma_2} \frac{f'(z)}{f(z)} dz$$

and deduce that $f(z)$ has just one zero inside the square $ABCD$.

14. By integrating $e^{ipz}/\cosh^2 z$ along the lines $\mathbf{I}(z) = 0$, $\mathbf{I}(z) = \pi$ prove that

$$\int_0^{\infty} \frac{\cos px}{\cosh^2 x} dx = \frac{\pi p}{2 \sinh \frac{1}{2}\pi p}.$$

15. By integrating

$$\frac{\operatorname{cosec} z}{(z-t)(z^2+1)}$$

round a square whose vertices are $(n+\frac{1}{2})(\pm\pi\pm i\pi)$ prove that

$$\frac{\operatorname{cosec} t}{t^2+1} + \frac{2et}{(e^2-1)(t^2+1)} = \sum_{n=-\infty}^{\infty} \frac{(-1)^n}{(1+n^2\pi^2)(t-n\pi)}.$$

16. By discussing

$$\int \frac{e^{iz\alpha}}{\cos \pi z} \frac{dz}{z-\zeta} \qquad (-\pi < \alpha < \pi)$$

round the circle $|z| = r$, where r is an integer, show that

$$\frac{\cos \alpha\zeta}{\cos \pi\zeta} = \frac{2}{\pi} \sum_{r=1}^{\infty} \frac{(-1)^{r+1}(r-\tfrac{1}{2}) \cos (r-\tfrac{1}{2})\alpha}{(r-\tfrac{1}{2})^2 - \zeta^2}.$$

17. By integrating $z^{-5} \sec \tfrac{1}{2}\pi z \operatorname{sech} \tfrac{1}{2}\pi z$ round the square $\mathbf{R}z = \pm 2N$, $\mathbf{I}z = \pm 2N$, where N is a positive integer, and making $N \to \infty$ prove that

$$\frac{\operatorname{sech} \dfrac{\pi}{2}}{1^5} - \frac{\operatorname{sech} \dfrac{3\pi}{2}}{3^5} + \frac{\operatorname{sech} \dfrac{5\pi}{2}}{5^5} - \ldots = \frac{\pi^5}{768}.$$

18. By considering

$$\int \frac{\coth \pi z \cot \pi z}{z^3} \, dz$$

taken round the square $x = \pm(N+\tfrac{1}{2})$, $y = \pm(N+\tfrac{1}{2})$ where N is a large positive integer, prove that

$$\sum_{1}^{\infty} \frac{\coth n\pi}{n^3} = \frac{7\pi^3}{180}.$$

INDEX

The numbers refer to the pages